Ore Deposits of the Gilman District, Eagle County, Colorado

By THOMAS S. LOVERING, OGDEN TWETO, *and* T. G. LOVERING

GEOLOGICAL SURVEY PROFESSIONAL PAPER 1017

Study made in cooperation with the Colorado Mining Industrial Development Board

Description of the ore deposits and geologic setting of a mining district that is the leading source of zinc in Colorado and an important source of silver, copper, lead, and gold

UNITED STATES GOVERNMENT PRINTING OFFICE, WASHINGTON : 1978

UNITED STATES DEPARTMENT OF THE INTERIOR

CECIL D. ANDRUS, *Secretary*

GEOLOGICAL SURVEY

H. William Menard, *Director*

Library of Congress Cataloging in Publication Data
Lovering, Thomas Seward, 1896–
Ore deposits of the Gilman District, Eagle County, Colorado.
(Geological Survey Professional Paper 1017)
Study made in cooperation with the Colorado Mining Industrial Board.
Bibliography: p. 83
Includes index.
Supt. of Docs. no.: I 19.16:1017
1. Ore-deposits–Colorado–Eagle Co.
I. Tweto, Ogden, 1912– joint author. II. Lovering, Tom Gray, 1921– joint author. III. Colorado Mining Industrial Development Board. V. Series:
United States Geological Survey Professional Paper 1017
TN24.C6L786 553'.09788 77-608257

For sale by the Superintendent of Documents, U.S. Government Printing Office
Washington, D.C. 20402
Stock Number 024-001-03098-5

Introduction

Sylvanite Publishing and Miningbooks.com are proud to put back into print this long out of print publication. The information in this publication is still valid and informative for historians, researchers, prospectors, miners, geologists and more. Much of this information is becoming lost as more and more books are discarded from libraries or schools, as older material disintegrates, and as publications are thrown out because someone does not know their value.

Our goal is to put hundreds of these publications back into print at a reasonable cost. Many of these originals can cost hundreds of dollars which is out of the reach of many who just want the information for study. In the process of this we try and clean up the books and text best we can to produce a quality product. There are many publishers reprinting books these days but they are essentially scan shops that put them into print errors, smudges, writing in the books, and all. Very occasionally we will do the same if we cannot obtain a copy in decent condition.

If you would like to see a particular book or are and author that has written a book in the past that is now out of print in our subject field, feel free to contact us and we will see what we can do to get that publication back in print.

CONTENTS

ILLUSTRATIONS

TABLES

SYSTEM OF MEASUREMENT UNITS

In the fieldwork on which this report is based, all measurements were made in English units. In the report, the metric equivalent of each English unit is shown in parentheses following the English unit. Conversion factors are listed below. In accord with convention, microscopic dimensions were measured and are recorded only in metric units.

English unit	*Multiplication factor*	*Metric unit*
miles (mi)	1.609	kilometers (km)
feet (ft)	.305	meters (m)
inches (in.)	2.54	centimeters (cm)
square miles (mi^2)	2.6	square kilometers (km^2)
pounds (lb)	.45	kilograms (kg)
atmosphere (atm)	1.013	bar

Metric unit	Multiplication factor	English unit
kilometers (km)	0.621	miles (mi)
meters (m)	3.281	feet (ft)
centimeters (cm)	.394	inches (in.)
millimeters (mm)	.039	inches (in.)
kilograms (kg)	2.2	pounds (lbs)
bar (kg/cm^2 or 10^5 Pa (Pascal)	.987	atmosphere (atm)

ORE DEPOSITS OF THE GILMAN DISTRICT, EAGLE COUNTY, COLORADO

By THOMAS S. LOVERING, OGDEN TWETO, and T. G. LOVERING

ABSTRACT

The Gilman mining district, known also in the past as the Red Cliff district, is in the mountains of southeastern Eagle County, west-central Colorado. The district is the leading source of zinc in Colorado and one of the major base-metal mining districts in the State. As valued at the time of production, total output of zinc, silver, copper, lead, and gold through 1972 was about $328 million. About 90 percent of this total was produced after 1930.

The productive part of the district is an area of about 3 square miles (7.8 square kilometers) on the northeast side of the deep canyon of the Eagle River between the small towns of Gilman and Red Cliff. The ore deposits are principally replacement deposits in dolomites of Mississippian and Devonian age and in quartzite of Cambrian age. A few productive veins occur in Precambrian rocks. The replacement deposits crop out in the cliffs of the canyon wall and extend northeastward downdip beneath Battle Mountain, which is composed of a thick sequence of Pennsylvanian clastic rocks. The deposits were originally worked through several separate mines along the canyon wall, but since 1918, all deposits in dolomite rocks, except some small ones near Red Cliff, have been worked through the Eagle mine of the New Jersey Zinc Company at Gilman.

The Gilman district lies on the eastern flank of the huge anticline of the Sawatch Range, near the steeply plunging north end of the anticline. Sedimentary rocks on the flank of this part of the anticline dip homoclinally northeastward to a synclinal axis about 8 mi (miles) (13 km (kilometers)) northeast of Gilman and then rise more steeply to the Gore fault at the edge of the Gore Range. The homocline is broken by only a few faults most of which have displacements of less than 100 ft (feet) (30 m (meters)). In contrast, the underlying Precambrian rocks are broken by numerous faults and shear zones related to the Homestake shear zone, a northeast-trending master shear zone several miles wide. Fractures and shear zones along the northwest side of the master zone extend beneath the Gilman district.

The Gilman district is at the northwestern edge of the northeast-trending Colorado mineral belt as defined by mineralized areas and bodies of intrusive porphyries. Neighboring mining districts in the mineral belt to the southeast of Gilman are the Kokomo lead-zinc district, 13 mi (21 km) distant; the Climax molybdenum district, 16 mi (26 km) distant; and the Leadville district, 20 mi (32 km) distant. These districts, as well as others farther away, are characterized by abundant intrusive rocks of Laramide and middle Tertiary ages and by complex faults systems. The Gilman district, in contrast, contains only a single sill of porphyry and has very simple geologic structure. This probably reflects a position either at the side of or high above a batholith that is inferred from geologic and geophysical data to underlie the mineral belt at shallow depth.

The rock column preserved in the district consists, in succession downward, of (1) grit, conglomerate, sandstone, and shale of the Pennsylvanian Minturn Formation, as much as 6,300 ft (1,920 m) thick; (2) shale, limestone, and sandstone of the Pennsylvanian Belden Formation, 200 ft (61 m) thick; (3) a sill of quartz latite porphyry of the Cretaceous Pando Porphyry about 80 ft (24 m) thick intruded in the basal shale of the Belden Formation; (4) dolomitized limestone of the Mississippian Leadville Dolomite, 110–140 ft (34–43 m) thick; (5) sandstone and sandy and cherty dolomite of the Mississippian or Devonian Gilman Sandstone, 15–50 ft (4.5–15 m) thick; (6) thin-bedded primary or diagenetic dolomite of the Mississippian(?) and Devonian Dyer Dolomite, 50–80 ft (15–24 m) thick; (7) quartzite, conglomerate, and green shale of the Devonian Parting Formation, 35–50 ft (11–15 m) thick; (8) sandstone and shale of the Ordovician Harding Sandstone, 14–80 ft (4.2–24 m) thick; (9) shaly and sandy dolomite and dolomitic glauconitic sandstone of the Cambrian Peerless Formation, 65–70 ft (20–21 m) thick; (10) white quartzite and dolomitic sandstone of the Cambrian Sawatch Quartzite, 190–220 ft (58–67 m) thick; and (11) Precambrian X Cross Creek Granite and associated migmatite.

Among the several stratigraphic units named, the Leadville Dolomite, Gilman Sandstone, Dyer Dolomite, Sawatch Quartzite, and the Precambrian rocks are the host rocks of ore deposits. The Leadville Dolomite, the principal host rock, is separated from the overlying Belden Formation by an unconformity that represents a period of karst erosion of the Leadville. Karst solution cavities within the Leadville are identified by the presence of a mixture of clay, silt, and chert fragments that is designated karst silt. Certain dissolution features in the Gilman Sandstone suggest that the karst circulation system extended in depth at least to the Gilman. However, the karst features of both the Gilman and the Leadville are obscured by dissolution features formed hydrothermally at the mineralization stage in Laramide time.

The sedimentary rocks of the district strike about N.45° W. and dip 10°–12° NE. Except for their homoclinal tilt, they are only slightly deformed. Folds occur only as very gentle warps that generally are apparent only upon mapping of a key bed. Despite their subtlety, the warps were a factor in the localization of hydrothermal processes. Like the fold structure, the fracture structure is also subtle. The fracture system consists of (1) bedding or low-angle faults or slip surfaces; (2) steep faults or "slips" of very small displacements and of limited length and limited vertical extent; (3) breccia zones in or almost in the plane of the bedding; and (4) joint systems that differ in orientation from place to place and also on opposite sides of bedding faults. All these features are interpreted as adjustments to regional folding.

Principal bedding faults are in the basal shale of the Belden Formation, in a bed known as the pink breccia in the upper part of the Leadville Dolomite, in the Harding Sandstone, in what is known as the Rocky Point zone near the top of the Sawatch Quartzite, and at the contact between the Sawatch and the Precambrian rocks. Many others are present locally throughout the pre-Belden stratigraphic sequence. Many of the high-angle slips or small faults in the sedimentary rocks either end upward and downward against bedding faults or bend abruptly into bedding faults. Movement on almost all the high-angle slips was parallel or subparallel to the bedding. Most faults in the Precambrian rocks terminate upward at the base of the Sawatch Quartzite and show low-angle strike-slip movement parallel to the base of the quartzite.

Rock alteration began with dolomitization of the Leadville Limestone, followed by recrystallization of parts of the dolomite rock thus formed. These processes operated before mineralization and throughout the width of the Colorado mineral belt; hence, they were not specifically related to mineralization at Gilman.

Alteration in dolomite rocks and quartzite at the mineralization stage consisted principally of dissolution. This process operated prior to and during deposition of sulfides. In the dolomite rocks, dissolution produced sanded or friable rock, loose dolomite sand, and open cavities or channelways. Many of the openings were filled with resedimented dolomite sand and cave rubble. The rubble locally contains fragments of porphyry from the sill in basal beds of the Belden Formation, indicating that the main period of dissolution postdated the porphyry. Dissolution was controlled in part by the ancient karst channel system and in part by fractures, such as the joint system and bedding-fault breccias. The Gilman Sandstone was intensely attacked and evidently was also a major element of the circulation system.

In the Sawatch Quartzite, dissolution was confined to the bedding-fault breccias and intersecting steep fractures in the Rocky Point zone near the top of the quartzite, principally in the area south of Gilman. The dissolution produced open cavities and channelways that are devoid of rubble.

Alteration along veins in the Precambrian rocks and in the porphyry sill above the ore bodies followed the sequence: argillic, sericitic, and late argillic.

The three minerals pyrite, siderite, and sphalerite constitute at least 90 percent of the total volume of ore deposits in the district. Galena is an important subsidiary mineral, and in certain ore bodies, chalcopyrite, tetrahedrite, and sulfosalt and telluride minerals are also important. Three different pyrites, designated I, II, and III, are recognized. Pyrite I is an early pyrite that was deposited monomineralically while dissolution was at its height. Pyrite II, the most abundant pyrite in the ore deposits in dolomite rocks, is associated with siderite and sphalerite. Pyrite III is a late, generally coarsely crystalline and auriferous pyrite present only in small quantities. As judged from texture, some pyrite II preexisted as marcasite. Marcasite is still preserved in some ore bodies in the Sawatch Quartzite. The siderite in the ore bodies in dolomite rocks typically contains about 10 percent of manganese and is also magnesian. Siderite of this same character is also the predominant type in certain ore deposits in the Sawatch Quartzite, though a paragenetically older manganese-poor variety is present locally. Typical sphalerite of the ore deposits is a black marmatite that contains an average of about 11 percent of iron. A very little iron-free sphalerite of younger age is present in some of the gold-, silver-, and copper-bearing ores.

The most productive ore bodies in the district are long subcylindrical replacement deposits that extend down the dip in the upper part of the Leadville Dolomite. These are referred to in the district as mantos. The four main mantos are 2,000–4,000 ft (610–1,220 m) long, 50–400 ft (15–122 m) wide, and 2–150 ft (0.6–46 m) thick. Each merges downdip with a chimney ore body that cuts across the bedding. Except at their oxidized upper ends, the mantos consist of pyrite, siderite, sphalerite, and subordinate, somewhat argentiferous galena. In most places the mantos have an outer shell that consists almost entirely of siderite, and at their lower or downdip ends they have an inner core that consists almost entirely of pyrite. The sulfide minerals and siderite of the mantos replaced bedded dolomite rock, dolomite sand in channelways, and the dolomitic material in cave rubble, forming, respectively, bedrock ore, sand ore, and rubble ore. Bodies of rubble and sand ores extend through the lengths of the mantos indicating that the pre-ore channel system (a joint product of karst and hydrothermal dissolution processes) was the immediate control for ore deposition. The mantos contain pockets of highly altered karst silt, especially in the lower or downdip parts, and they are capped in many places by post-ore collapse breccia.

The second most productive group of ore bodies consists of the chimenys at the lower ends of the mantos. The chimneys are funnel-shaped bodies that extend downward from the top of the Leadville Dolomite through the Gilman Sandstone and Dyer Dolomite to terminations at or in the Parting Formation or Harding Sandstone. The height of the chimneys to the base of the Dyer is about 240 ft (73 m), and the largest chimney is about 400 ft (122 m) in diameter at the top. The chimneys are composite ore bodies that consist of two quite different aggregates of materials. The older aggregate in each chimney is geometrically the root of the connected manto. Before remineralization, each chimney consisted of a core of massive pyrite (pyrites I and II) surrounded by a shell of black sphalerite ore, and this, in turn, was surrounded by an outer shell of siderite. At some time after the chimney ore bodies of this type were created, parts of the pyrite cores of the chimneys were irregularly remineralized with a suite of silver- and gold-bearing copper and lead minerals, forming what is called copper-silver ore. The upper parts of the chimneys, in the Leadville Dolomite, consist principally of rubble and sand ores, and they contain irregular bodies both of altered, clayey karst silt and of post-ore collapse breccia derived from the shale and porphyry cap rocks. The lower parts of the chimneys, in the Dyer Dolomite, consist principally of bedrock ore.

Ore deposits in the Sawatch Quartzite have been far less productive than the mantos and chimneys in dolomite rocks but were a major source of gold and silver in the early years of mining in the district. Most of the production was made from the oxidized zone, which in most of the mines extends downdip about 1,000 ft (305 m) from the outcrop. In the unoxidized deposits, quartzite beds and fractures in the Rocky Point zone near the top of the quartzite are widely mineralized with pyrite I, which contains only traces of silver, gold, and copper except where secondarily enriched. The pyrite is closely related in distribution to solution channels, and some of it fills or lines the channels. Within the broad pyritized areas are sharply defined belts oriented in the dip direction that are overprinted with deposits characterized by siderite, pyrite II or marcasite, and small amounts of black sphalerite, chalcopyrite, and galena. These deposits occur as coatings or fillings in cavities and fractures in either fresh or pyritized quartzite. In one mine — the Bleak House — they form a zinc-lead manto replacement deposit astraddle a vein that occupies a small fault. In other mines, the sideritic deposits have been worked only in the zone of secondary enrichment. Auriferous pyrite III occurs in veinlets and small pockets in scattered localities in the quartzite, principally in the Ground Hog mine. Pockets of telluride minerals occur with this pyrite in a few places. In the intensely leached upper part of the oxidized zone, gold and the silver chloride mineral cerargyrite occur as granules and nuggets in a pasty argillic bedding seam. Deeper in the oxidized zone, these granules and also gold in an invisible form occur in small pipes or thin ribbonlike seams of iron oxides and sulfates within large bodies of nearly barren iron oxides.

Scattered northeast-trending veins are present in the Precambrian rocks beneath the mineralized sedimentary rocks. Though a few have been moderately productive, the veins as a group have contributed only a fraction of one percent to the total output of the district. The most productive veins occupy faults of presumed Laramide age within belts of sheared rocks that constitute a part of the Precambrian Homestake shear zone. Most of the veins terminate upward at the base of the Sawatch Quartzite; no physical connection between the veins and the ore deposits in the upper Sawatch and the dolomite rocks is known. The veins have been mined chiefly for their precious metal content. They consist mainly of pyrite and quartz but contain chalcopyrite, black sphalerite, and galena in scattered small ore shoots. A little late pyrite, which is inferred to be pyrite III and is accompanied by late, resinous sphalerite, is present locally.

Small telluride deposits are present in a few localities in basal beds of the Sawatch Quartzite above veins in the Precambrian rocks. The telluride minerals occur as tiny veinlets or films on random joints in glassy quartzite and are unaccompanied by any other mineral. The veinlets consist principally of hessite but also contain petzite, which is veined by native gold.

A small mineralized center at Red Cliff seems to be distinct from the main center at Gilman. Deposits in the Red Cliff area were worked in several small mines, mainly in the early years of mining in the area. The deposits are in the upper part of the Leadville Dolomite. The ore bodies are small and irregular replacement deposits along minor faults in brecciated dolomite rock. Most of the ore mined was either argentiferous lead carbonate ore or partly oxidized and secondarily enriched pyrite-sphalerite-galena ore. The high silver content of the ore together with local occurrences of gold and bismuth suggest alliance with the copper-silver ores in the chimneys at Gilman.

Collectively, the various ore deposits indicate two mineralization stages — an early and main stage characterized by pyrite, siderite, and black sphalerite and a later stage characterized by copper and precious metals. The early stage is inferred to have occurred in Laramide — specifically, Paleocene — time, though definitive age data are lacking. The late stage may have followed after only a brief pause, or it might have occurred much later, possibly during the Oligocene episode of mineralization in the mineral belt nearby, as at Climax.

The rock cover over the Leadville Dolomite at the early mineralization stage at Gilman is estimated to have been 12,000 ft (3,660 m) thick. The cover thickened to the northeast from Gilman and thinned to the southwest; accordingly, a gradient of decreasing pressure to the southwest existed. If the late mineralization stage occurred as late as Oligocene time, the cover at Gilman may have been as little as 8,000 ft (2,440 m) thick, and the pressure gradient may have been steep.

Dolomitization and mineralization were accomplished by quite different hydrothermal processes. We relate dolomitization to the interplay of three factors or events: (1) the presence in the region of a thick saline-water-bearing sequence of Pennsylvanian and Permian clastic rocks; (2) the rise of the anticline of the Sawatch Range at the beginning of Laramide orogeny, causing the saline waters to start circulating; and (3) gradual regional heating above a batholith that is inferred to have begun its rise beneath the mineral belt with the onset of Laramide orogeny. Isotopic data suggest that dolomitization of the Leadville was accomplished in the temperature range 150°–175°C. The inferred temperature due to geothermal gradient is about 145°C; thus, only slight magmatic heating was required to cause the initial dolomitization. With continued heating, the dolomite recrystallized in many areas in the mineral belt. Isotopic and limited fluid-inclusion data suggest that, at Gilman, the temperature reached 300°C when zebra rock formed. Dolomitization probably began about 72 million years ago, and recrystallization probably continued to the beginning of the mineralization stage, about 60 million years ago.

Mineralization began with the introduction of solutions that were highly corrosive toward both dolomite and quartzite and that deposited pyrite. These solutions moved only in restricted courses through a rock body saturated with the saline waters in which dolomite had recrystallized. Changes in rates of flow and routes of travel allowed the ambient ground waters to close in at times, causing temperature reversals, such as that implied by the presence of marcasite. The solutions entered the Gilman area from the downdip (northeast) direction and at first moved principally through the fractured rocks of the Rocky Point zone of the Sawatch Quartzite. Slightly later, they broke through the overlying sedimentary rocks at the sites of the chimney ore bodies and reached the karst channel system in the Leadville. Hence, the flow in the Sawatch Quartzite was greatly reduced at the time of major sulfide deposition in the dolomite rocks.

The ultimate source of the solutions is not known but is presumed to have been in a crystallizing magma body east or southeast of Gilman. Whatever the source, the early solutions must have moved through conduits insulated by altered zones in the Precambrian rocks in order to arrive in the sedimentary rocks in such a potent state. The marked contrast between the ores of the early and late mineralization stages suggests either a change in source or a marked change at a single source.

INTRODUCTION

The Gilman mining district, in the mountains of southeastern Eagle County (fig. 1), is the leading source of zinc in Colorado and one of the major base-metal mining districts of the State. Ore deposits were discovered in the district in 1879 and were worked continuously from that time to the present (1974). On the basis of value at the time of production, total output of zinc, silver, copper, lead, and gold from the district reached about $328 million at the end of 1972. About 90 percent of this total was produced after 1930, considerably later than the periods of peak production from many other Colorado mining districts.

The ore deposits of the Gilman district are principally replacement deposits in dolomites of Mississippian and Devonian age; replacement deposits of much smaller value occur in underlying quartzites of Cambrian age, and relatively minor vein deposits occur both in Precambrian basement rocks and in the pre-Pennsylvanian Paleozoic sedimentary rocks. The deposits are generally similar in composition and stratigraphic occurrence to those of some other mining districts in west-central Colorado, notably those of the Leadville district, 20 mi (32 km) to the south, and of the Aspen district, 32 mi (51.5 km) to the southwest (fig. 1). However, the Gilman district lacks the complex fault patterns that characterize the other two districts, and it lacks the complex pattern of igneous intrusion that is also a characterizing feature of the Leadville district. In short, the Gilman district is distinctive for the outward simplicity of its geologic setting. The features that controlled or affected ore deposition are subtle, as will be shown.

INVESTIGATION AND REPORTS

The ore deposits of the Gilman district were studied by us principally in the early 1940's in conjunction with the geologic mapping of the Minturn quadrangle. The geology of this quadrangle is discussed in a separate

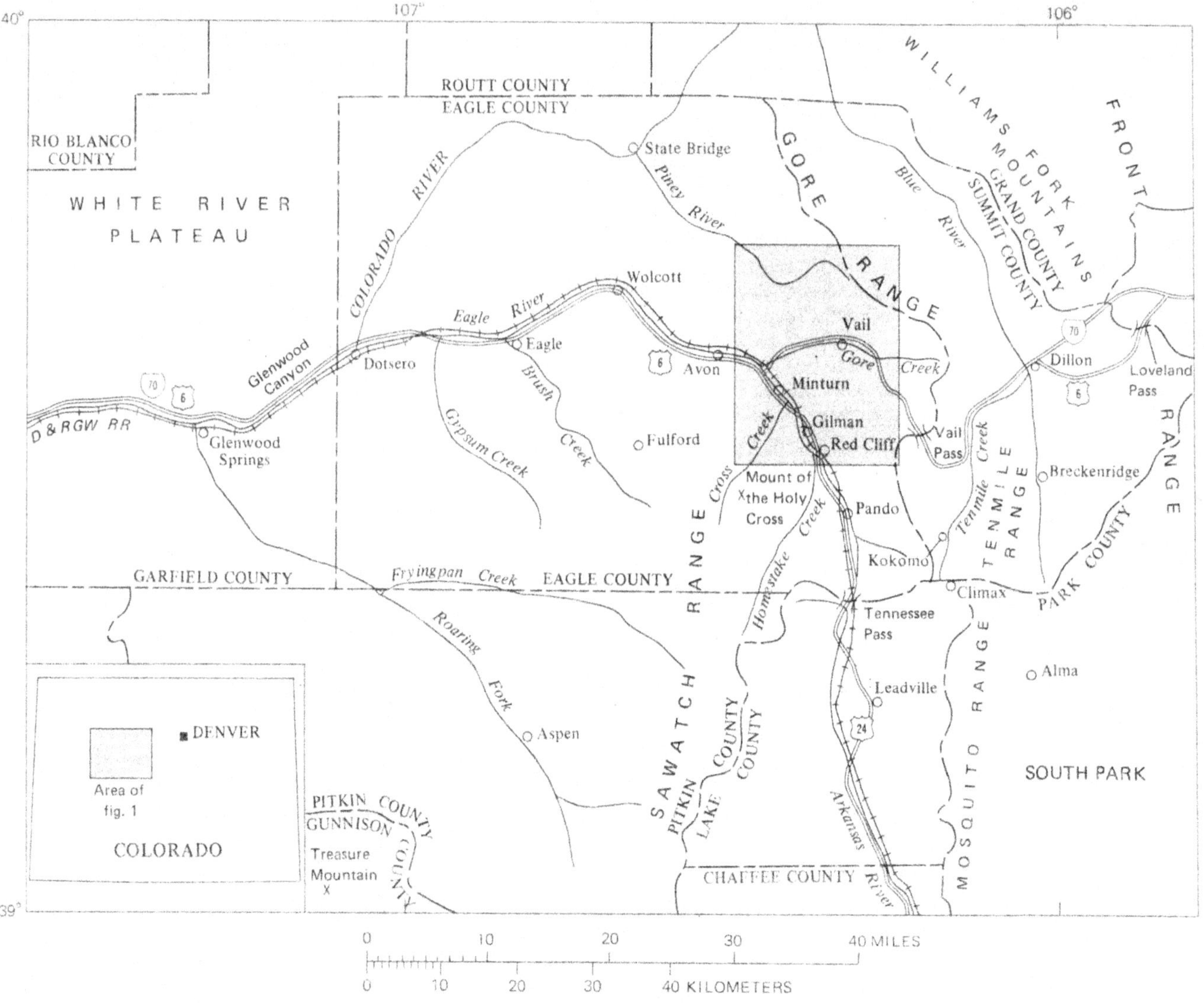

FIGURE 1. — Index map of west-central Colorado. The Gilman district is small area between Gilman and Red Cliff, within Minturn 15-minute quadrangle (shaded).

report (Tweto and Lovering, 1977), to which the reader is referred for more extended discussions of many geologic features than will appear in this report. Principal results of our studies of the ore deposits were summarized in two early reports (Lovering and Tweto, 1944; Tweto and Lovering, 1947), but for reasons detailed in the quadrangle report, publication of a final summary has been long delayed. In the interim, an excellent factual report on the ore deposits of the district by geologists of the New Jersey Zinc Co., owner of most of the district, was published (Radabaugh and others, 1968). In addition, numerous reports on special topics related to the ore deposits — noted as appropriate in the text that follows — have also appeared since the time of our chief work in the district. Our aim at this late date is in part to integrate all these studies, but mainly to document concepts introduced in our preliminary reports of 1944 and 1947, to discuss features and interpretations that we regard as controversial, and, particularly, to relate the district as an altered and mineralized center to its geologic setting and history.

To accomplish these objectives, we shall first present the purely factual data on the geology, rock alteration, mineralogy, and ore deposits of the district; then we will discuss our inferences as to the geologic and chemical processes that together produced a very remarkable concentration of ore deposits. In this latter endeavor we are admittedly hampered by our failure to utilize the many advanced techniques of investigation that have become available since the time of our early work in the district. We explain this omission on two grounds: (1)

our observations, and particularly our mineral and rock sampling, were based on the investigatory techniques and the concepts of geologic processes that existed prior to World War II, and (2) any attempt at this late date to upgrade our data to the standards of the 1970's would further delay a report that is already unconscionably delayed.

So that the reader will have a basis for judgment, we note here that our data on ore minerals are based on polished-section studies and etch tests under the microscope, supplemented by a few spectroscopic analyses and a very few X-ray analyses. Most of the work on the ore minerals was done by Tom G. Lovering in the mid-1950's. The rock-alteration studies — including those of clays — were made microscopically, principally by Thomas S. Lovering.

PHYSICAL SETTING

The productive part of the Gilman district occupies an area of about 3 mi^2 (7.8 km^2) bounded approximately by the Eagle River and Rock, Willow, and Turkey Creeks (pl. 1). The Eagle River in this part of its course flows in a deep and precipitous canyon. The Denver and Rio Grande Western Railway follows the river in the canyon bottom. U.S. Highway 24 follows a course above the canyon rim, across the steep slopes of Battle Mountain. Gilman, a "company town" of the New Jersey Zinc Co. and the site of the principal mining operations during the last several decades, is perched on a narrow promontory at the rim of the canyon, 600 ft (183 m) above the river (fig. 2). The town of Red Cliff, 2 mi (3.2 km) southeast of Gilman, is in the canyon bottom at the mouth of Turkey Creek. Prior to the establishment of Gilman in about 1918, Red Cliff was the only town in the area, and the mining district was generally called the Red Cliff district. The district was also known as the Battle Mountain district, the name used in early land records.

The ore deposits of the Gilman district crop out in the cliffs of the canyon wall and extend northeastward downdip beneath the slopes of Battle Mountain. Early-day mines were located at several sites along the canyon wall between Gilman and Red Cliff. Later, most of these mines were consolidated under one ownership, and the ore deposits were worked principally through a shaft at Gilman and an adit near the level of the canyon bottom at the railroad siding called Belden.

FIGURE 2. — Canyon of the Eagle River at Belden (in canyon bottom) and Gilman (at top of cliffs). Cliffs of stratified rock in middle part of canyon wall are Sawatch Quartzite, which lies on Precambrian rocks. Discontinuous cliffs higher on canyon wall are Chaffee Group and Leadville Dolomite. Vertical distance between Belden and Gilman is about 600 ft (183 m).

GEOLOGIC SETTING

REGIONAL FEATURES

The Gilman district is on the eastern flank of the huge anticline of the Sawatch Range and very near the north end of the anticline, which plunges steeply northward (fig. 3). West of the canyon of the Eagle River, sedimentary rocks on the flank of the anticline have been stripped down to the resistant lower Paleozoic formations, which form dip slopes that extend high onto the flank of the range. Glaciated canyons cut through the sedimentary formations of the dip slopes deep into Precambrian basement rocks. East of the canyon of the Eagle River, thousands of feet of Pennsylvanian rocks, constituting the Belden and Minturn Formations, are preserved above the lower Paleozoic formations (fig. 4). These rocks form Battle Mountain and the mountains eastward for 10 mi (16 km) to the Gore fault (fig. 3), the major bounding fault of the Gore Range (Tweto and Lovering, 1977, pl. 1).

Though the sedimentary rocks of the Gilman district are broken only by minor faults, the underlying Precambrian rocks are broken by numerous faults and shear zones. These features are elements of a major northeast-trending Precambrian shear zone — the Homestake shear zone — the main part of which passes to the south of the Gilman district (Tweto and Sims, 1963; Tweto, 1974). This master shear zone, which consists of numerous individual shear zones and faults in a belt 6–7 mi (9.7–11.3 km) wide, extends through the Precambrian terrane of the Sawatch Range and disappears northeastward beneath the flanking sedimentary rocks. The shear zone has a distinctive magnetic expression even where buried, and its course beneath the sedimentary terrane to the Gore fault is plainly evident on a magnetic map (Tweto and Case, 1972, pl. 2). At the Gore fault, the compound shear zone narrows to a single prominent shear zone which continues northeastward across the Precambrian terrane of the Gore Range (Tweto and others, 1970, pl. 1).

The part of the Homestake shear zone present in the Gilman district is significant for two reasons: (1) most of the veins in the Precambrian rocks occupy fractures of the shear zone system that were reopened in later (presumably Laramide) time, and (2) the shear zone may have been a part of the fundamental plumbing system for the Gilman ore deposits, as discussed in following sections.

The Gilman district is at the northwestern edge of the Colorado mineral belt as generally defined (fig. 3). This belt is characterized by mineralized centers and by porphyritic igneous intrusive bodies of Late Cretaceous and Tertiary age. In addition to the Leadville and Gilman districts, the segment of the mineral belt between the Mosquito and Sawatch Ranges contains the Kokomo lead-zinc district, about 13 mi (21 km) southeast of Gilman, and the Climax molybdenum deposit, 16 mi (25.9 km) southeast of Gilman. The Leadville, Climax, and Kokomo districts are alined along the generally north trending Mosquito fault, the western border fault of the Mosquito and Tenmile Ranges (fig. 3).

The intrusive igneous rocks of the mineral belt consist principally of two age groups, one Late Cretaceous and early Tertiary (Laramide) and one Oligocene. In a few places, members of a third group, of late Miocene age, are also present. Igneous bodies of the two main age groups are interspersed through much of the length of the mineral belt, though the Oligocene group predominates west of the Sawatch Range. The intrusive rocks, in the form of dikes, sills, stocks, and laccoliths, have long been regarded as manifestations of an underlying batholith (Crawford, 1924; Lovering and Goddard, 1938). Gravity data in the region between the Sawatch and Mosquito Ranges strongly support this inference (Tweto and Case, 1972).

The Gilman area, which contains only one igneous body — a sill — is a few miles north of the complexly intruded part of the mineral belt. The edge of the area of abundant and varied intrusive rocks is near Pando, about 5 mi (8 km) south of Red Cliff (Tweto, 1953; 1974). A northwest-trending line of small stocks in the Precambrian rocks of the Sawatch Range lies about 7 mi (11.3 km) southwest of Gilman. This line terminates to the northwest with a moderately large stock at Fulford, 13 mi (21 km) west of Gilman (fig. 3). The stocks of this line support the inference (Tweto, 1975) that the Sawatch anticline is more an elongate intrusive dome athwart the trend of the mineral belt than a purely tectonic feature.

Mineralization and rock alteration are associated with all three of the intrusive stages noted previously. For example, the Kokomo lead-zinc deposits, which occur as replacement deposits in limestone beds of the Pennsylvanian Minturn Formation, were linked by Bergendahl and Koschmann (1971) with quartz monzonite intrusions that are known from their geologic relations to dated igneous rocks (Pearson and others, 1962) to be of Laramide age. In contrast, the Climax molybdenum deposit, only 3 mi (4.8 km) from Kokomo, is genetically related to an Oligocene granitic stock (Wallace and others, 1968). Very weakly expressed mineralization features in the Gore Range, north of Kokomo and northeast of Gilman, have been inferred on geologic grounds to be of late Miocene and younger age (Tweto and others, 1970). West of the

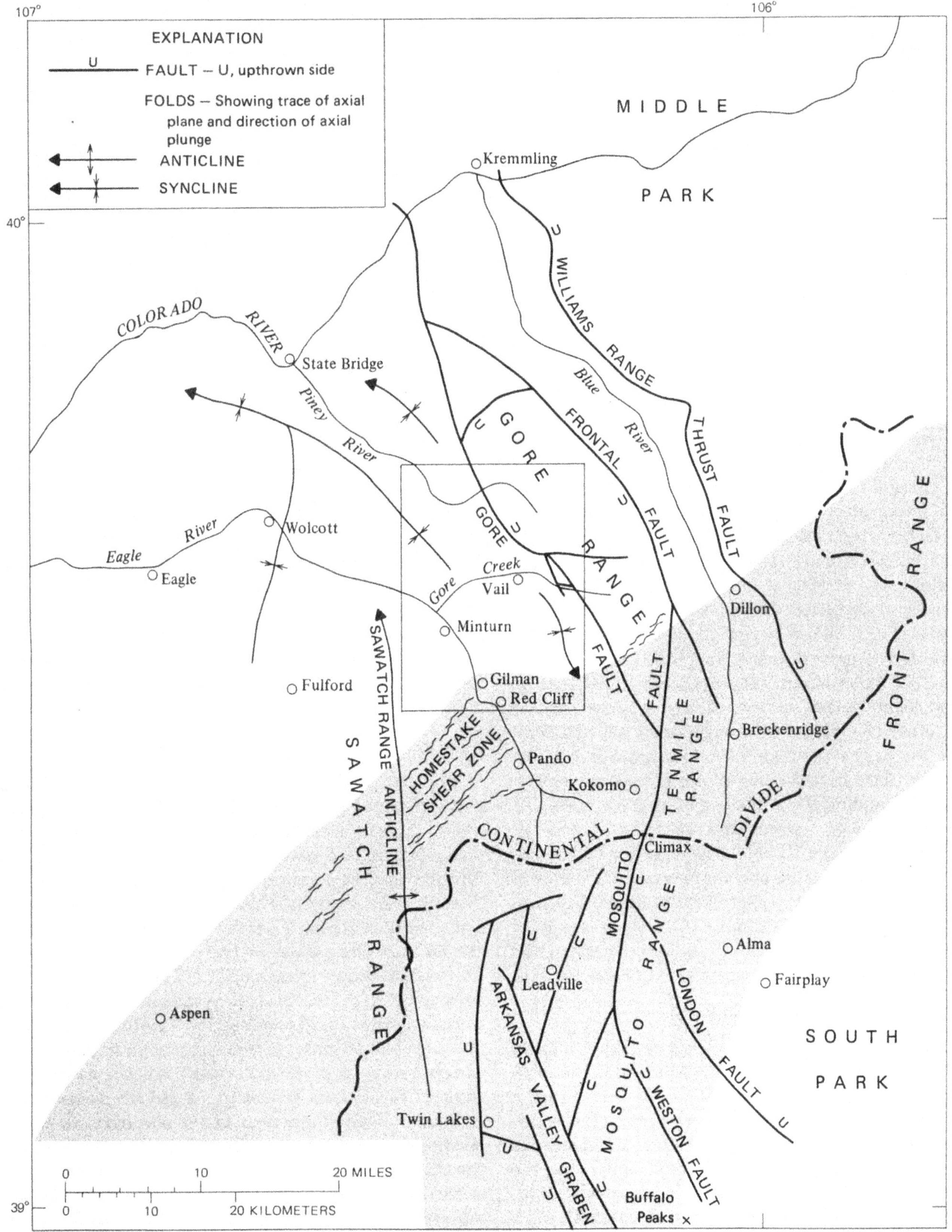

FIGURE 3. — Major structural features in region surrounding Gilman and the Minturn 15-minute quadrangle (outlined), and approximate area of the Colorado mineral belt (shaded).

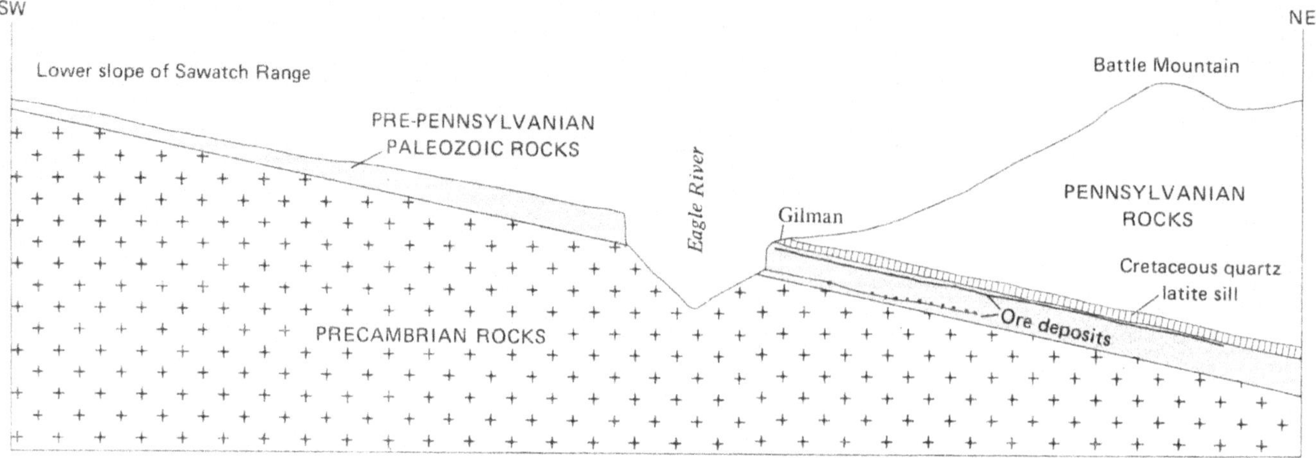

FIGURE 4. — Diagrammatic cross section of the Gilman area. Length of section is about 4 mi (6.5 km); vertical scale is slightly exaggerated.

Sawatch Range, at Treasure Mountain (Vanderwilt, 1937) mineral deposits are associated with a granite stock dated as 12.4 m.y. (million years) (late Miocene) in age (Mutschler, 1970).

At Gilman and also throughout the neighboring part of the mineral belt, the Mississippian Leadville Limestone, the chief host rock of the replacement ore deposits is dolomitized. Moreover, successive facies of recrystallized dolomite recognized at Gilman are also distributed widely through the belt. Replacement facies, such as jasperoid, and dissolution facies, such as sanded dolomite, that occur at Gilman are found also in many other localities in the neighboring mineral belt.

All the features of geologic and igneous history, mineralization, and rock alteration in the area bordering or surrounding the Gilman district have a bearing on the origin, age, and localization of the Gilman ore deposits. Clearly, the Gilman district is not an isolated entity independent of its surroundings, but it is an element of a much larger pattern and problem. Our purpose in this report is to examine the district closely as one of the important elements of the larger pattern and to study the relationship of the district to the pattern.

STRATIGRAPHY

Rock units present in the Gilman district are listed in table 1 and are described in the Minturn quadrangle report (Tweto and Lovering, 1977). We here discuss only the stratigraphic features pertinent to the occurrence and origin of the ore deposits and attendant rock alteration. Because economic interest centers on the Leadville Dolomite and decreases progressively downward to the Precambrian rocks, the discussion begins with the top unit of table 1, and the intrusive sill is treated as a rock layer in the sequence. Altered facies of many of the units are discussed further in the section on rock alteration.

Though rock units above the Minturn Formation are no longer preserved in the district, some part of an additional sequence several thousand feet ($\approx$2 km) thick was inferentially present at the time of mineralization, and an even thicker sequence was present earlier. As described in part in the Minturn quadrangle report (Tweto and Lovering, 1977), the vanished sequence consisted, in succession upward, of the Pennsylvanian and Permian Maroon Formation, the Triassic Chinle Formation, the Jurassic Entrada and Morrison Formations, and the Cretaceous Dakota, Benton, Niobrara, and Pierre Formations.

MINTURN AND BELDEN FORMATIONS

The Minturn and Belden Formations, of Middle Pennsylvanian age, constitute a thick cap of predominantly clastic rocks over the mineralized carbonate rocks. As exposed on the slopes of Battle Mountain, the Minturn and Belden are unmineralized and unaltered. They provide essentially no indications of the rich ore bodies beneath them. Thin dolomite beds or pods within the Minturn Formation on the top of Battle Mountain are geochemically anomalous in silver and show some evidence of hydrothermal recrystallization, but no more so than similar carbonate beds of the Minturn in many places in the Minturn quadrangle and adjoining quadrangles. As seen in the mine workings, the basal black shale of the Belden Formation is plastically deformed as a result both of intrusion of the porphyry sill and of bedding-fault movements. In addition, in many places the shale has sagged or collapsed into channels or caves formed in the underlying Leadville Dolomite during the mineralization process, as discussed in a later section.

Though barren, the Minturn and Belden Formations are significant for their influence on some of the hydrothermal processes that affected the underlying rocks. For one thing, the shaly Belden Formation constituted a barrier of low permeability that obviously was a factor in guiding the circulation of mineralizing and altering fluids. More fundamentally, the Pennsylvanian rocks of these formations or their equivalents are inferred to have been an important factor in the dolomitizing process that preceded mineralization. We here discuss briefly some regional features of these rocks as background for later discussions of processes.

About 2 mi (3.2 km) northwest of Minturn (fig. 1), the Minturn Formation gives way by facies change to evaporitic rocks now classed as the Eagle Valley Evaporite or Formation (Lovering and Mallory, 1962; Bartleson and others, 1968). The evaporitic facies, characterized principally by gypsum or anhydrite but containing salt in places, extends downward into the Belden Formation and also extends upward above the Minturn Formation into the overlying Maroon Formation. The evaporitic facies is preserved in a broad arc enclosing the northern end of the Sawatch Range (Mallory, 1971) and presumably once existed over at least this part of the range. To the southeast of the main body of evaporitic rocks, in the region from Minturn to South Park (fig. 1), the Belden and Minturn contain scattered lenses of gypsum or anhydrite and, in places, abundant salt casts. Similarly, evaporitic materials occur among the dominantly clastic Pennsylvanian rocks on the west side of the Sawatch Range southward to a latitude 17 mi (27.5 km) south of Aspen (Bruce Bryant, oral commun., 1974). The Minturn and Belden Formations almost certainly extended across the site of the Sawatch Range before that range was elevated, despite suggestions to the contrary by some authors (De Voto, 1972, p. 159). With their content of evaporitic materials and before disturbance in Laramide time, these units constituted an immense reservoir of saline connate waters. The role of such waters in the processes leading to mineralization is discussed in a closing section.

PANDO PORPHYRY

The intrusive sill of the Gilman area is a fine-grained quartz latite porphyry assigned to the Pando Porphyry, the oldest and most widespread of all the porphyries in the mineral belt south and southeast of Gilman. Sills of

TABLE 1.—*Rock column in the Gilman district*

Age	Unit	Thickness, feet (meters)	Character
Middle Pennsylvanian	Minturn Formation	6,300 (1,921)	Arkosic grit, sandstone, conglomerate, and shale; some intercalated dolomite and limestone.
	Belden Formation	200 (61)	Black shale, sandstone, and limestone. Very thin lenses of regolithic Molas Formation (Pennsylvanian) locally present at base.
Late Cretaceous	Intrusive sill of Pando Porphyry	80 (24)	Altered fine-grained quartz latite porphyry. Sill is in basal beds of Belden.
	UNCONFORMITY		
Early Mississippian	Leadville Dolomite	110–140 (34–43)	Upper unit: massive light-gray extensively recrystallized dolomite, uneven in thickness; Lower unit: bedded fine-grained dark-gray dolomite and medium-grained gray dolomite; 90–95 ft (27–29 m) thick; cherty in lower part.
	UNCONFORMITY		
Early Mississippian or Late Devonian	Chaffee Group: Gilman Sandstone	15–50 (4.5–15)	Sandstone, sandy and cherty dolomite, lithographic dolomite, and breccia; extensively modified by alteration in mineralized area.
	UNCONFORMITY		
Early Mississippian(?) and Late Devonian	Chaffee Group: Dyer Dolomite	50–80 (15–24)	Thin-bedded fine-grained gray dolomite.
Late Devonian	Chaffee Group: Parting Formation	35–50 (11–15)	White to tan quartzite, quartz pebble conglomerate, and green shale.
	UNCONFORMITY		
Middle Ordovician	Harding Sandstone	14–80 (4.2–24)	Argillaceous sandstone, quartzite, and green clay shale.
	UNCONFORMITY		
Late Cambrian	Peerless Formation	65–70 (20–21)	Shaly and sandy dolomite and dolomitic glauconitic sandstone.
	Sawatch Quartzite	190–220 (58–67)	Medium- to thick-bedded medium-grained white quartzite; dolomitic sandstone in local beds in upper part and in small lenses in middle part.
	UNCONFORMITY		
Precambrian X	Cross Creek Granite		Coarsely porphyritic gray to pink granite; contains lenses of migmatite or schist and is cut by belts of cataclastic gneiss.

this porphyry occur within the Belden Formation throughout the area from Gilman to Leadville and beyond, and in much of the region they occur also in other formations from the Sawatch through the Minturn (Tweto, 1974). In almost all occurrences, the porphyry is extensively altered deuterically, and in many places a later hydrothermal alteration is superposed on the deuteric alteration. Fresh Pando Porphyry from one of the source plutons near Leadville has been dated by the K-Ar method as 70 m.y., or late Late Cretaceous, in age (Pearson and others, 1962).

In the mine area at Gilman, the porphyry sill is about 80 ft (24 m) thick; from there, it thickens gradually to the south and thins to north. The sill is in the basal strata of the Belden Formation, generally 5–15 ft (1.5–4.6 m) above the Leadville Dolomite. The sill disturbed the adjoining shale but had no metamorphic effect on the shale. On the contrary, the shale probably contributed to the deuteric alteration of the porphyry, as suggested by differences in the character of this alteration with differences in wallrocks (Tweto and Lovering, 1977). The deuterically altered porphyry is further altered hydrothermally in the vicinity of the ore bodies, as discussed in the section on rock alteration. In many places, porphyry from the sill has collapsed along with rocks from the Belden Formation into caverns or channels cut into the Leadville Dolomite during mineralization. Where collapsed material was replaced by sulfides, the porphyry fragments were largely unaffected. Thus, they remain as fragments suspended in sulfide ore, and they indicate clearly that metallization postdated porphyry intrusion.

PRE-BELDEN UNCONFORMITY

The unconformity between the Belden Formation and the Leadville Dolomite represents a rather lengthy time interval from about the middle of the Mississippian Period to the middle of the Pennsylvanian Period. Events during this interval are unknown except that during the latter part of the interval a karst erosion surface was formed on the carbonate rocks of the Leadville. The karst surface is irregular on both regional and local scales. On a regional scale, it accounts for marked differences in the thickness of the Leadville from area to area. On a local scale, the irregularities take the form of pinnacles and crevices at the top surface of the Leadville and of channels and caves in the carbonate rock beneath this surface. In the cliffs just below U.S. Highway 24 at the curve 0.3 mi (0.5 km) northwest of Gilman (pl. 1), for example, the top surface of the Leadville has a relief of as much as 15 ft (4.5 m) in a horizontal distance of 10 ft (3 m).

In many places a regolithic deposit, consisting in part of the residue from the chemical weathering of the carbonate rocks and in part of silt or very fine quartz sand from other sources, fills the depressions in the top surface of the Leadville or forms a discrete layer over this surface. Where such a layer is recognizable as a thin stratigraphic unit, it is designated as the Molas Formation. In the Gilman area, the chief occurrence of the regolithic material is as a filling in caves, channels, and crevices within the Leadville Dolomite; material that might be referred to the Molas Formation in a stratigraphic sense is present only locally, in small pockets or thin lenses. In the text that follows, we shall refer to the regolithic material introduced into cavities in the Leadville in pre-Belden—presumably Pennsylvanian—time as karst silt. As will be seen, another system of cavities and cavity-fills developed much later, at the mineralization stage, and it is important to distinguish the two.

In the mine workings, the karst silt is observed principally in or near the sulfide ore bodies. From this, it is inferred (1) that the karst channel system was erratically distributed through the area and was strongly developed only locally, and (2) that the karst system influenced the circulation of fluids at the mineralization stage. The karst silt in or close to the ore bodies is intensely altered and is a white to light-green pasty and plastic material containing abundant pieces of chert. Most of the chert is fragmental, but some is in discoidal or ellipsoidal bodies 1–10 in. (2.5–25 cm) in diameter whose surfaces are characteristically marked by smooth, more or less concentric corrugations. It is suspected, though not proved, that these chert bodies grew in the silt, but whatever their origin, they are a characteristic feature of the karst silt and Molas Formation throughout the region.

Most of the karst silt bodies in the mine workings have been so distorted by plastic flow caused both by bedding-fault movements and by disturbance created by mine openings that their original form and sizes are indeterminate. They are normally by-passed or lagged-off in mining, and the chief visual evidence of them is as a white paste extruded through cracks in the lagging. Some of the silt bodies have apparent volumes on the order of cubic yards or tens of cubic yards, but some enclosed in the ore bodies must have volumes in the hundreds of cubic yards.

Mineralogy of the altered karst silt (long called "shaly lime" by the mine staff at Gilman) is discussed in the section on alteration.

LEADVILLE DOLOMITE

The Leadville Dolomite of Early Mississippian age is the principal host rock of ore deposits in the Gilman district, just as it is in the Leadville, Aspen, and several other districts in the Sawatch and Mosquito Ranges. Through most of its area of occurrence in Colorado, this unit consists principally of limestone and is known as the Leadville Limestone. Through the width of the mineral belt, however, from near Gilman to Buffalo Peaks (fig. 3), a distance of more than 40 mi (65 km), the unit consists entirely of dolomite and is referred to as the Leadville Dolomite. The northwestern boundary between the limestone and dolomite facies is at the edge of the Gilman area, at Cross Creek.

The Leadville of the Gilman district consists of two main lithic units. The upper unit of the Leadville is a massive light-gray coarse-grained extensively recrystallized dolomite known to the mine geologists as the "discontinuous banded zone." This unit is of variable thickness, owing to the unevenness of the karst erosion surface at its top, but typically it is 40–50 ft (12.2–15.2 m) thick. The lower unit, 80–100 ft (24.4–30.5 m) thick, consists of dark-gray to black finely crystalline dolomite and subordinate interbedded medium-gray and medium-crystalline dolomite. The lower 50–65 ft (15.2–19.8 m) of the unit contains abundant black chert. The two units of the Leadville are separated by a distinctive brecciated and slightly sandy and shaly dolomite zone, typically 1–4 ft (0.3–1.2 m) thick, that is known as the pink breccia. This zone marks a minor unconformity that has been modified by bedding-fault movement and hydrothermal alteration. The two units recognized in the dolomite facies of the Leadville correspond in general to two lithic units widely recognized in the limestone facies outside the mineral belt. As described by several authors (Engel and others, 1958; Banks, 1967; Nadeau, 1972; Conley, 1972), the Leadville Limestone typically consists of two units: an upper, massive bioclastic and oolitic limestone and a lower, bedded fine-grained to lithographic gray limestone. The two units are separated by an unconformity, in many places obscure, that presumably correlates with that of the pink breccia zone.

We interpret the dolomite facies of the Leadville as an early hydrothermal feature produced by replacement of limestone during the initial stages of Laramide tectonism and magmatism. The products of this process and various later recrystallizations are discussed in the section on rock alteration.

GILMAN SANDSTONE

The Gilman Sandstone of Early Mississippian or Late Devonian age evidently was an important factor in the circulation of mineralizing and rock-altering fluids. Thus, this thin unit has a significance out of proportion to either its stratigraphic magnitude or its minor role as a host rock of ore deposits. In the Gilman area, just as at Leadville and other mineralized centers, the Gilman Sandstone has been considerably modified by the fluids that circulated in it. Consequently, it shows many departures in lithology, contacts, and thickness from the norm elsewhere.

Where not appreciably altered, the Gilman averages about 20 ft (6 m) in thickness and consists of sandstone, dolomite, chert, and breccias of these materials. The sandstone is concentrated in the lower part of the formation, generally in beds 8–20 in. (20–51 cm) thick. The sandstone is medium to coarse grained, consists of rounded and frosted quartz sand grains, and is friable to firmly cemented. The cement is mainly carbonate but in a few places is silica. The dolomite is fine grained to lithographic, is finely laminated to structureless, and closely resembles the dolomite of the underlying Dyer Dolomite. Except in a top stratum of the formation, much of the dolomite is sprinkled with quartz sand and contains small lenses of dolomitic sandstone. The top unit of the formation is a very fine grained structureless dolomite known locally as the waxy bed. Unlike the dolomite of the Leadville, that of the Gilman is concluded to be of primary or diagenetic origin, on the basis of isotopic composition of its oxygen (Engel and others, 1958) and sedimentational features (Banks, 1967; Nadeau, 1972). The breccias in the Gilman are erratically distributed, contributing appreciably to the overall heterogeneity of the formation. The breccias are presumably of sedimentary origin; the mechanisms of origin were discussed by Banks (1967) and Nadeau (1972). The Gilman is bounded above and below by smooth-surfaced unconformities of low relief (Tweto and Lovering, 1977). In at least one locality in the mine workings, however, the Gilman fills a broad channel cut to a depth of about 30 ft (9 m) in the underlying Dyer Dolomite.

In most parts of the mineralized area, the Gilman differs markedly from the unit just described. In those places, it is characterized by abundant black clay, by internal slump and collapse structures, by very irregular contacts with the Dyer below and the Leadville above, and by features that indicate disaggregation and resedimentation of sandstone and dolomite. All these features are discussed further in the section on rock alteration.

DYER DOLOMITE

The Dyer Dolomite of Early Mississippian(?) and Late Devonian age is a fine-grained gray to dark-gray thin-bedded dolomite that exhibits many indications of shallow-water deposition. Wavy bedding surfaces coated with argillaceous matter and lenses of intraformational dolomite breccias are common in the formation. Thinly laminated, generally crenulated beds suggest stromatolitic origin (Campbell, 1970). A thin but persistent stratum known as the "sand grain marker" contains round frosted quartz sand similar to that in the Gilman Sandstone. This stratum, 45 ft (13.7 m) above the base of the Dyer, is a particularly useful datum in assessing displacements on the many slip surfaces in the rocks and in defining the gentle warps that, as discussed later, were a factor in the localization of ore bodies. The Dyer is a primary or diagenetic dolomite (Campbell, 1970) which except for dissolution was affected only to a minor degree by the alteration processes that affected the Leadville.

In the Gilman district, the Dyer Dolomite is a host rock only of chimney ore bodies that extend downward from the Leadville Dolomite. In the Leadville district, by contrast, the Dyer is a major host rock of ore deposits (Tweto, 1968a). In areas where carbonate rocks have been extensively replaced by jasperoid, the Dyer was replaced on an equal or greater scale than the Leadville. Such areas exist on the northwest fringe of the Gilman district (Tweto and Lovering, 1977, pl. 1) and in the Pando area, south of Red Cliff (Tweto, 1953).

PARTING FORMATION AND HARDING SANDSTONE

The Parting Formation of Late Devonian age, which consists of quartzite and subordinate green shale, and the Harding Sandstone of Middle Ordovician age, which consists of green sandstone and shale, were evidently null components in the alteration and mineralization processes. They are mineralized only to a minor degree at the bottom of ore chimneys, and the shale in them is bleached and altered only in the most intensely mineralized areas. The unconformity between the Upper Devonian Parting Formation and the Middle Ordovician Harding Sandstone was, however, a locus of major low-angle fault movements. The unconformity at the base of the Harding exists in lieu of the Lower Ordovician Manitou Dolomite. The Manitou, which is widespread in the mineral belt south of the Gilman district (Tweto and Lovering, 1977), is an important host rock of ore deposits at Leadville, where it was known in the older literature as the "White limestone." In this respect, the Gilman district contains one less potential ore horizon than do the Leadville and other districts in this part of the mineral belt.

PEERLESS FORMATION AND SAWATCH QUARTZITE

Basal beds of the Peerless Formation and the underlying upper part of the Sawatch Quartzite, both formations of Upper Cambrian age, constitute a significant ore zone in the Gilman district, though far subordinate to the dolomite formations in importance. Ore deposits in this zone are principally in the Sawatch; they extend upward into the Peerless only locally. The mineralized zone in the upper Sawatch is known locally as the Rocky Point zone. At the town of Gilman, in the Rocky Point and Bleak House mines (fig. 11), the Rocky Point zone is characterized both by distinctive lithology and by brecciation caused by bedding- or low-angle-fault movements. In this area, the zone consists of 1 to 10 ft (0.3–3 m) of Peerless-type sandy dolomite and dolomitic sandstone interbedded in the vitreous white quartzite of the Sawatch at a level 10–15 feet (3–4.6 m) below the top of the Sawatch. Farther south, in the Percy Chester, Ground Hog, and Champion mines (fig. 11), dolomitic rocks are absent in the Sawatch, and the Rocky Point zone consists of one to three brecciated beds or bedding zones at various stratigraphic levels in the upper 20–30 ft (6–9 m) of the Sawatch.

The rocks in some parts of the Rocky Point zone were extensively altered hydrothermally, as discussed in the section on alteration. Other than in this zone, the Sawatch is unaltered and barren except for minor occurrences of silver-and-gold telluride veinlets and thin pyritic replacement deposits in the basal beds. Basal dolomitic and glauconitic sandstones of the Peerless Formation are mineralized locally in the Bleak House and Champion mines. Shaly dolomite of the upper Peerless is altered and is somewhat mineralized beneath some of the ore chimneys in the overlying dolomite formations.

PRECAMBRIAN ROCKS

The Precambrian X rocks beneath the Gilman district are mainly coarse-grained porphyritic Cross Creek Granite, described in the Minturn quadrangle report (Tweto and Lovering, 1977). The granite locally contains lenses or screens of biotite gneiss or migmatite, and westward from the canyon of the Eagle River, it grades rapidly into migmatite. Along shear zones of the Homestake shear zone system, as in the Ben Butler and Tip Top mines (fig. 11), the granite and included gneisses are reduced to a dark-greenish-gray cataclastic gneiss in belts as much as 50 ft (15.2 m) wide. Hydrothermally altered facies of the Precambrian rocks are discussed in the section on rock alteration.

STRUCTURE

The Gilman district is part of a uniform homocline tilted northeastward on the flank of the Sawatch anticline. Sedimentary rocks of the homocline on the average strike N. 40°–45° W. and dip 10°–12° NE. Warps and fractures within the homocline — as seen in the pre-Belden sedimentary rocks — are on a very small scale but are significant to the occurrence of the Gilman ore deposits. Warps or distortions of the strata are of two kinds: one on fairly large scale and tectonic in origin and one local, of shrinkage origin, near caves, collapse structures, and ore bodies. Among the tectonic warps, one of the more evident is a structural terrace in which the strata flatten from the normal dip of 10°–12° to 6°–8°. The terrace strikes northwest and is about 200 ft (60 m) wide. A line of ore chimneys is near, and parallel to, this terrace. Very gentle anticlinal and synclinal warps oriented either in the strike or the dip directions of the homocline are also present in the area exposed by mine workings. Most of these warps are so gentle as to be unevident to the eye; they are evident only upon mapping of a marker horizon at some fixed altitudinal plane. Despite their subtlety, warps of this type were factors in the localization of ore bodies, as discussed in a later section.

The fracture system of the homocline consists of (1) bedding or low-angle faults and slip surfaces; (2) steep faults or "slips" of small displacement and limited length; (3) breccia zones, mainly in the plane of the bedding; and (4) joint systems of various trends and degrees of prominence, depending upon location. All these features are closely interrelated. We interpret them as products of adjustment to the rise of the Sawatch anticline in early Laramide time.

LOW-ANGLE AND BEDDING FAULTS

Low-angle or bedding faults occur throughout the stratigraphic section and are a characteristic structural feature of the Gilman district. The most persistent zones of low-angle-fault movements are in the basal strata of the Belden Formation, in the pink breccia zone of the Leadville Dolomite, in the Harding Sandstone, in the Rocky Point zone of the Sawatch Quartzite, and at the contact between the Sawatch Quartzite and the Precambrian rocks. In addition, bedding-slip surfaces are evident locally throughout the pre-Belden stratigraphic sequence exposed in the mine workings. Many of these slips are localized at the thin shaly partings scattered rather abundantly throughout the dolomite and quartzite formations. (See stratigraphic sections, Tweto and Lovering, 1977.)

The bedding faults in shaly zones, such as those of the Harding Sandstone and the Belden Formation, display drag folding and are inferred to have had relatively large, though indeterminate, movements. Movements in the persistent breccia zones, such as the pink breccia of the Leadville Dolomite and the Rocky Point zone of the Sawatch Quartzite, probably were of smaller magnitude. The breccia of the Rocky Point zone is principally a crackle breccia; this breccia is bounded in places by vertical slips that show very little displacement. The breccia of the pink breccia is in part a crackle breccia and in part a disorganized breccia. The movements on some of the local bedding or low-angle faults amounted at least to a few tens of feet (several meters), as indicated by displacements on the faults that are not exactly parallel to the bedding. Apparent displacements on many other bedding slips were only of the magnitude of centimeters or millimeters, as determined from offsets of steep features, such as carbonate veinlets. However, movements may have occurred before such features formed, as well as after.

The directions of bedding or low-angle fault movements were varied and changed with time, as determined from close studies in several parts of the mine workings. Drag folds in the Harding indicate that the principal movement was of a reverse-fault type — that is, the strata above the movement zone moved updip to the west or southwest with respect to the strata below. Such movement is consistent with adjustment to the rise of an anticline. A fairly persistent movement zone at the top of the Parting Formation also shows reverse-fault movement, as determined from offsets coupled with the orientation of striae. In the Rocky Point zone, at least two stages and directions of movement are evident. Slip surfaces within the zone indicate that the upper beds first moved northward with respect to the lower ones, and then, after or during fracturing that produced northeast-trending joints, they moved northeast, downdip. A movement zone about 25 ft (7.6 m) above the base of the Dyer Dolomite shows early strike-slip movement that carried the upper beds northwest and a later normal-fault movement in which these beds moved due east. Both normal- and reverse-fault movement occurred in the Belden Formation, as indicated by the orientation of drag folds in different places. We have no data on the sequence of these movements.

Faults or fractures of the bedding-fault orientation are not confined to the sedimentary sequence but occur also in the underlying Precambrian rocks. Prominent joints or slips oriented parallel to the sedimentary rocks lend a stratified appearance to the granitic rocks in parts of the walls of the canyon of the Eagle River. As seen in mine workings that extend horizontally through Precambrian rocks to the base of the Sawatch Quartzite, such fractures are most abundant near the base of the quartzite and decrease in number and prominence with depth below the quartzite.

HIGH-ANGLE FAULTS

Mapped faults of the Gilman district are shown on plate 1. In discussion that follows, we shall also consider a group of faults of smaller magnitude that are known principally from the mine workings.

Faults of the area (pl. 1) are not numerous and have

only small displacements as measured in the sedimentary rocks. Most of them trend northeast, and many in the Precambrian rocks do not extend into the sedimentary rocks. Among those that do extend into the sedimentary rocks, several die out upward and do not reach the top of the Leadville Dolomite. Observations in mine workings in the Precambrian rocks indicate that movements on the faults were primarily of the strike-slip type. Grooves and striations on the walls of most of the faults plunge northeast at low angles, approximately parallel to the dip of the overlying sedimentary rocks. In accord with this orientation, horizontal displacements far exceed the vertical. One of the veins in the Ben Butler mine (pl. 2*E*) displaces a pegmatite dike 135 ft (41.1 m) in the horizontal plane but displaces the base of the Sawatch Quartzite only 40 in. (1 m) in the vertical plane. Another vein in this mine displaces a pegmatite dike 45 ft (13.7 m) horizontally but does not extend into the quartzite. The vein terminates upward in the gouge of a bedding fault at the base of the quartzite, and the basal surface of the quartzite above the gouge is an unbroken plane (fig. 5).

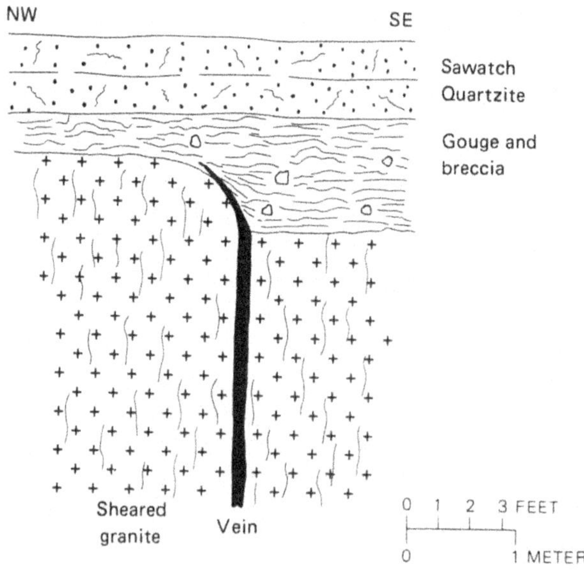

FIGURE 5.—Vein termination at base of Sawatch Quartzite in Ben Butler mine. Vertical and horizontal scales are equal.

The presence of a bedding fault at the base of the quartzite suggests the possibility that the steep faults and veins might be offset by the bedding fault rather than terminating against this fault. However, no indications of such offset have been found either at the surface or in mine workings. Moreover, some veins or faults extend for a short distance into the quartzite and displace it slightly. Examples are the vein in the Ben Butler mentioned previously and the nearby Star of the West vein (pl. 2*E*), which shows a vertical separation of about 5 ft (1.5 m) at the base of the quartzite. These relationships indicate that the bedding-fault movement preceded or accompanied, but did not follow, the movements on the high-angle faults or vein fissures. Therefore, the bedding fault cannot have displaced the high-angle fractures.

As a group, the northeast-trending faults coincide in location with a major northern strand of the Homestake shear zone (fig. 3). This strand of the shear zone transects the Precambrian rocks of the Sawatch Range several miles southwest of Gilman, disappears northeastward beneath the sedimentary rocks of the dip slopes west of the Eagle River, reappears in the canyon in the Gilman area, and projects northeastward beneath the sedimentary rocks east of the canyon (Tweto and Lovering, 1977, pl. 1; Tweto and Sims, 1963, pl. 1). A line of discontinuous small faults in the Minturn Formation northeast of the Gilman district coincides with the projection of the shear zone; this suggests that the shear zone may extend to the Gore fault northeast of Vail (Tweto and Lovering, 1977, pl. 1).

We regard most of the northeast-trending faults of the Gilman area (pl. 1) as fractures of the Precambrian shear zone system that were reactivated in Laramide time. A few may not have formed until Laramide time. The vein fissures of the Ben Butler, Tip Top, and Santa Cruz mines (fig. 11; pl. 2*E*) are simple fractures or narrow brecciated and sheeted zones within broad belts of cataclastically foliated rocks of the same general trend as the fissures. The fissures displace pegmatite and aplite dikes that, in part at least, seem to be younger than the cataclasis. The Star of the West vein shows additional relationships (fig. 6). In the Precambrian rocks, this vein (or fault) consists of breccia and gouge in a zone as much as 10 ft (3 m) wide between walls of sheared, cataclastically foliated granite. In the Sawatch Quartzite, in contrast, the fault consists only of a few tight fractures in a zone 2 ft (60 cm) wide. The marked

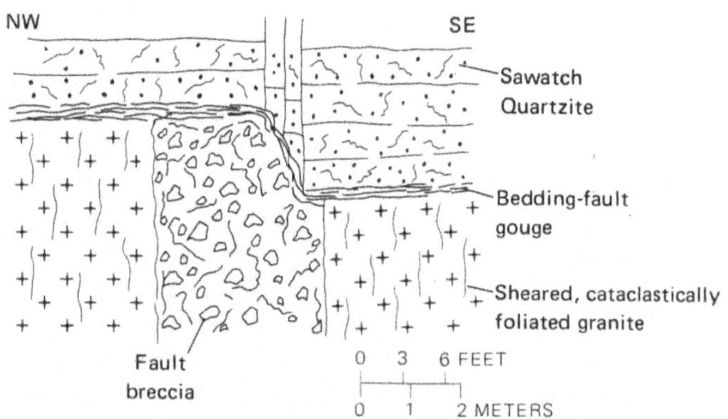

FIGURE 6.—Fault relations in Star of the West mine, lower incline. Vertical and horizontal scales are equal.

difference in character of the fault in the two rocks strongly suggests the existence of a fault in the Precambrian rocks before the fault in the quartzite developed, and probably before the Sawatch Quartzite was deposited.

TONGUE FAULTS

Mine workings in the sedimentary rocks expose many small high-angle faults or slip surfaces that are limited vertically to plates of rock between bedding faults and that, in general, are short or nonpersistent along strike. Most of these faults trend northeast, but some trend northwest. Almost all are strike-slip faults; grooves and striae on them plunge approximately parallel to the bedding. The faults either terminate upward or downward against bedding faults, or they bend into the bedding faults in a smooth but tight curve. We refer to such faults as tongue faults because they form the sides of prisms of rock that moved as independent tectonic tongues in the plane of the bedding (fig. 7).

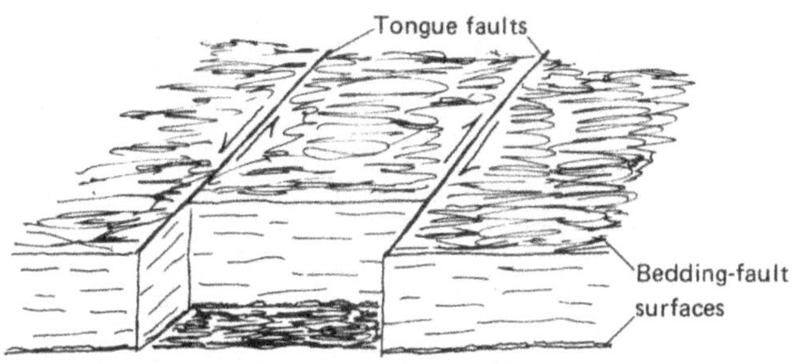

FIGURE 7.—Tongue faults. Scales are variable.

Most tongue faults show little or no vertical separation of beds on their two sides, but where the bounding bedding faults deviate from exact parallelism with the bedding, or where bedding is locally distorted, they do show vertical separation. Magnitude of the strike-slip movements on the tongue faults generally can be determined only where a vertical separation is evident. The magnitudes vary widely. One of the largest faults in the mine workings in sedimentary rocks is the Venus fault (fig. 45), a tongue fault that extends through the Dyer Dolomite and part of the Leadville Dolomite. The fault trends northeast and is nearly vertical. Striae on the walls have an average plunge of about 5° NE., and bedding on the two sides of the fault dips 10° NE. A vertical separation of 4 ft (1.2 m) evident along the fault indicates a strike-slip movement of about 45 ft (13.7 m). Traced upward, the fault bends abruptly and splays into a group of closely spaced bedding faults. The lower terminus is not exposed in the mine workings, nor is the fault evident in workings a few hundred feet northeast of the area of exposure. The movements on many other tongue faults were much smaller than those observed on the Venus fault. The relation of tongue faults to sulfide occurrence at a locality in the Sawatch Quartzite is illustrated in figure 46.

Tongue-fault and bedding-fault movements were evidently the cause of the extensive brecciation in the Rocky Point zone of the Sawatch Quartzite. As seen in mine workings (pl. 2*A*–*D*), this zone is characterized by a network of nonpersistent high-angle fractures or slips, as well as by brecciated beds. Bodies of crackle breccia in an individual quartzite bed are also nonpersistent and are generally bounded by the high-angle slips. In many places different beds of quartzite are brecciated on the two sides of a high-angle slip that produced no vertical separation of bedding planes, or only a few centimeters of separation (fig. 8). Some of the high-angle slips in turn are observed to terminate upward or downward against unbroken beds of quartzite. Others extend upward and downward beyond the limits of observation in the mine workings; though some of these have observed strike lengths of as much as 1,000 ft (305 m), they display little or no vertical separation of quartzite beds. The two most persistent fractures known in the mines in the Sawatch Quartzite are the Bleak House vein beneath Gilman and the Cleveland fault in the Ground Hog mine (fig. 50; pl 2*A*). The Bleak House vein has a vertical separation of about 2 ft (60 cm) and has been followed in mine workings at least 3,000 ft (915 m). The Cleveland fault shows a vertical separation of about 3 ft (1 m) and has been explored through a length of 1,300 ft (396 m).

The strike-slip faults in the Precambrian rocks underwent the same type of movement as the tongue faults in the sedimentary rocks, and at least their later movements were undoubtedly a response to the same forces.

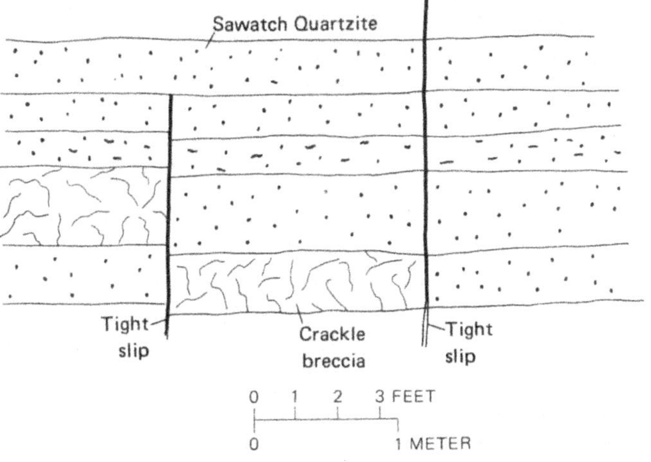

FIGURE 8.—Relations of brecciated beds and steep slips in Rocky Point zone of Sawatch Quartzite. Vertical and horizontal scales are equal.

IMPLICATIONS OF FRACTURE PATTERN

The fracture system of the Gilman district is consistent with adjustments to regional folds, and except for the element of inherited structure in the basement rocks, we believe it to be of such an origin. The fine structure indicates that the adjustment occurred by small differential movements between local prisms of rock. In its subtlety, the fracture system of the district differs markedly from those of neighboring mineralized centers in the mineral belt. This has a bearing on any speculations regarding the source of the ore deposits.

All nearby mining districts — Leadville, Kokomo, Climax, Breckenridge, Alma, and Aspen — are characterized by complex, "shattered-glass" fault patterns. These patterns are associated with the occurrence of abundant igneous intrusive rocks, and they almost certainly resulted in some part from the movements of magma at shallow depth beneath the surface now exposed to view. Additionally, the Leadville, Climax, and Kokomo districts are in the broad fracture zone of a major mountain-border fault — the Mosquito fault. This fault, one of the greatest in the region, is inferred to have guided porphyry magmas and attendant ore deposits into the upper crust (Tweto, 1968a,b). Moreover, the Mosquito and associated faults were reactivated in middle and late Tertiary time to serve as elements of the basin-and-range fault system that created the major Arkansas Valley graben in the flank of the Sawatch anticline from the latitude of Leadville southward (Tweto, 1968a; Tweto and Case, 1972). Given this tectonic environment, the fracture patterns and even the ore deposits of the nearby districts are understandable. Gilman, in contrast, displays no history of intense intrusion and fracturing, yet it contains major ore deposits of the same general character as those of many of its neighbors. The district demonstrates that even in a province with the magmatic signature of the Colorado mineral belt, major ore deposits can exist in the absence of trappings such as visible igneous bodies and complex fault structure.

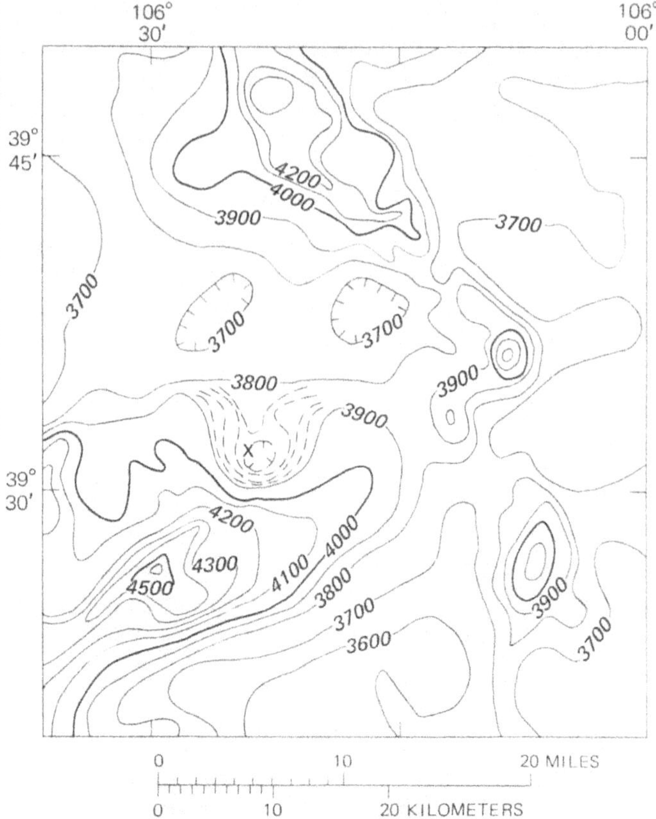

FIGURE 9. — Aeromagnetic map of region surrounding Gilman, after Zietz and Kirby (1972). Contours indicate total intensity magnetic field, in gammas, relative to arbitrary datum. Contour interval 100 gammas, with supplementary contours (dashed lines) near Gilman (X). Ticks on depression contours. Line of magnetic highs and high magnetic gradients trending northeastward from southwest corner of map marks Homestake shear zone. Northwest-trending line of magnetic highs in north-central part of map marks Gore Range.

MAGNETICS AND GRAVITY

The Gilman district lies within a small but sharply defined magnetic low (fig. 9). The low is a lobe of a larger east-trending low to the north, and the lobe is deeply incised into a major northeast-trending magnetic high to the south. The low at Gilman is defined by at least six east and west flight lines (U.S. Geological Survey, 1968), and, therefore, it is thought to be real and not a magnetic shadow.

The cause of the magnetic low is not readily evident and can only be speculated upon. Whatever the cause, the source must lie in the Precambrian basement rocks, as the sedimentary rocks are effectively magnetically transparent. The broad east-west low to the north of the Gilman area corresponds in general with the locations of large bodies of Cross Creek Granite in the Gore and Sawatch Ranges and is inferred to express the magnetic contrast between this granite and bordering, strongly magnetic migmatites (Tweto and others, 1970). The Precambrian rock at Gilman is principally Cross Creek Granite, and this gives way to migmatite immediately to the west and south (Tweto and Lovering, 1977, pl. 1). Perhaps, then, the magnetic low at Gilman only reflects the presence of a small lobe of granite protruding into a gneiss terrane. Alternatively, the low might express an altered area in which the magnetite of the Precambrian rocks has been destroyed. This possibility is not supported by the facts insofar as observations can be made. The Precambrian rocks in the canyon at Gilman are altered only in narrow sheaths enclosing the veins or faults. Through the length of the canyon, at least 95 percent of the Precambrian rocks are not visibly altered. Thus, if alteration is the cause of the magnetic low, it must be in rocks that are not exposed to view, yet not at great depth. No mine workings or boreholes in the Precambrian rocks extend to depths below the level of the Eagle River. As still another possibility, the magnetic low might be caused by a hidden intrusive body, such as a Laramide or Tertiary stock. The absence of intrusive rocks, except for the sill at the base of the Belden Formation that intruded from sources near Leadville, argues against this possibility. However, stocks can exist in the absence of appendages or satellitic bodies. Some of the small Laramide stocks in the Precambrian rocks of the Sawatch Range southwest of Gilman display no evidence of off-shooting or satellitic intrusions, such as dikes. These stocks, obviously now seen at a comparatively great crustal depth, are basically concordant in the Precambrian gneisses and intertongue with the gneisses in lit-par-lit arrangements. If such a stock exists beneath the Gilman area, its magnetic properties would be hard to predict. Some stocks of the mineral belt are associated with magnetic highs, some are associated with lows, and some display no magnetic contrast with their environment (Tweto and Case, 1972).

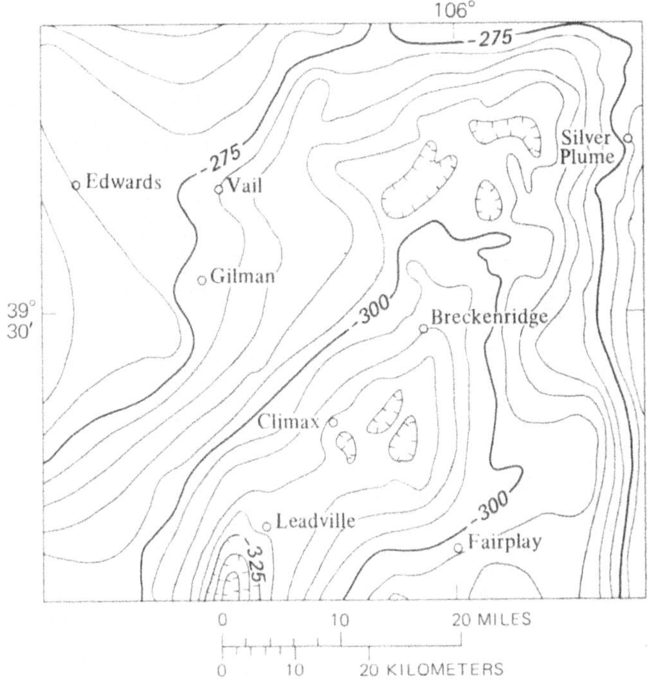

FIGURE 10.—Bouguer gravity map of region surrounding Gilman, after Behrendt and Bajwa (1974). Contour interval 5 milligals; ticks on depression contours. Northeast-trending "gravity valley" outlined by contours coincides approximately with the Colorado mineral belt and is interpreted to indicate an underlying batholith.

In its gravity setting, the Gilman district is on or beyond the northwest fringe of the "gravity valley" that characterizes the mineral belt (fig. 10). This gravity feature is interpreted to reflect an underlying batholith (Case, 1965; Tweto and Case, 1972). High gravity gradients along the sides of the gravity valley imply that the top of the batholith may lie as little as a few thousand feet (1–3 km) below the surface in the area from Breckenridge southwestward beyond Leadville (Tweto and Case, 1972, p. 20). The Gilman district is located in an area of lower gradient, near the southeast end of a gentle northwest-trending gravity ridge (fig. 10). Thus, as judged from the gravity data, either the Gilman area is some distance northwest of the inferred batholith or the batholith is at much greater depth beneath Gilman than it is in the Breckenridge-Climax-Leadville area. In this respect as in others noted previously, the Gilman district differs in setting from neighboring mineralized centers.

DISTRIBUTION AND TYPES OF ORE DEPOSITS

The chief source of the metals produced from the Gilman district is a group of large connected ore bodies beneath Battle Mountain at Gilman (fig. 11). These ore bodies are principally in the Leadville Dolomite, though some extend downward into or through the Dyer Dolomite. Ore deposits exist also at other localities and in other stratigraphic settings. On the basis of composition, form, and stratigraphic setting, six major groups or kinds of ore deposits are recognized in the district.

The most productive group of ore bodies consists of long subcylindrical replacement deposits extending downdip in the upper part of the Leadville Dolomite. These ore bodies are called mantos by the mine staff (Radabaugh and others, 1968), and we follow this established terminology, though technically the ore bodies are not the blanket deposits implied by the term "manto." The four main mantos, 2,000–4,000 ft (610–1,220 m) long, converge downdip and are united at their lower ends by a manto with a northwest-trending strike (fig. 11). Except at their oxidized upper ends, the mantos consist principally of pyrite, sphalerite, and siderite. They are valuable chiefly for zinc but contain some lead and small amounts of silver and copper. The parts of the mantos that contain substantial amounts of sphalerite are classed as zinc ore by the mine staff, and parts that consist almost entirely of pyrite — mainly at the lower ends of the mantos — are classed as "noncommercial iron." The mantos and also the other types of deposits noted here are discussed in detail in following sections.

The second most productive group of ore deposits

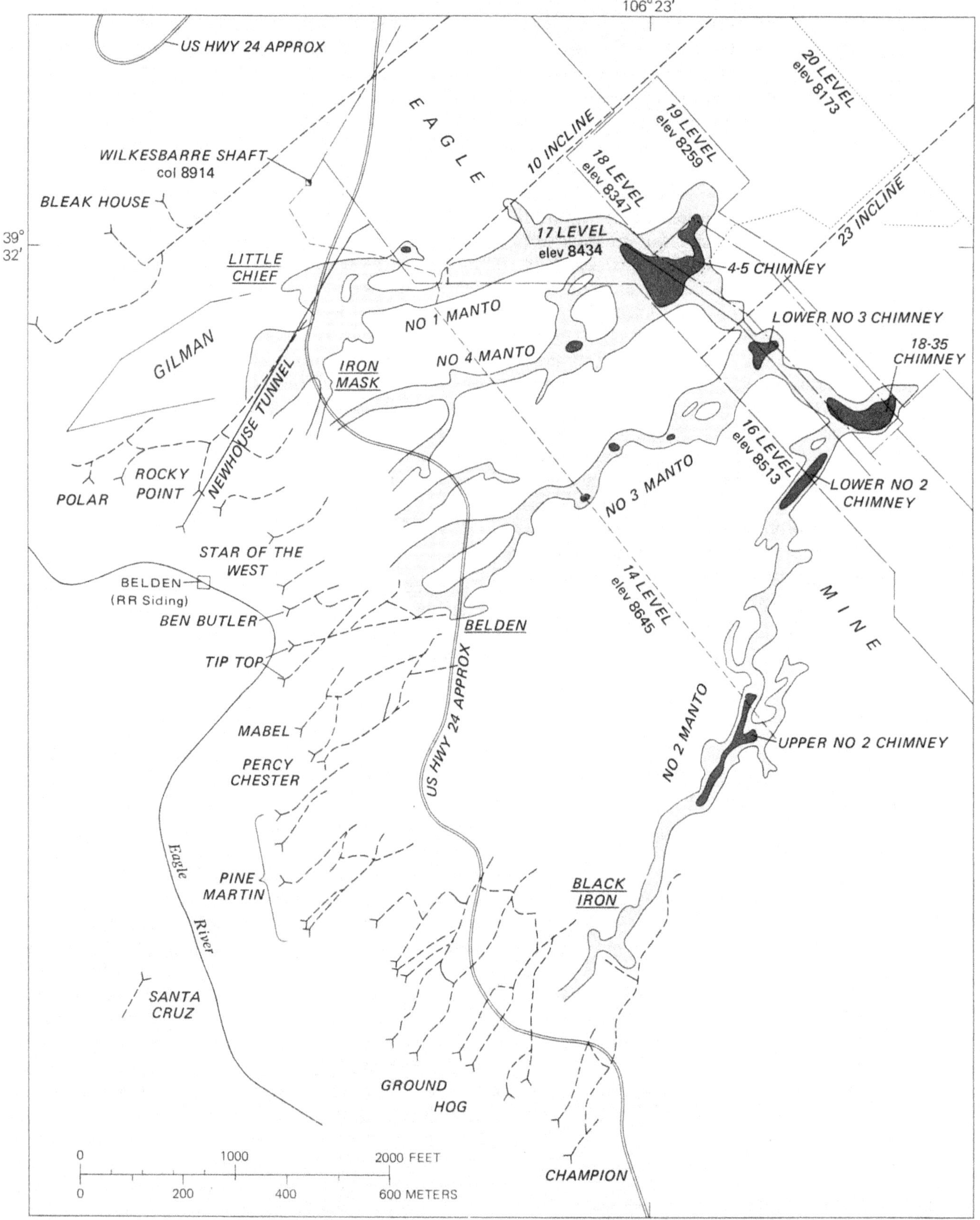

FIGURE 11. — Ore bodies and key workings of the Eagle mine, general location of chief mines (underlined) in Leadville Dolomite before consolidation into the Eagle mine, and locations and generalized outlines of main workings of mines in Sawatch Quartzite and Precambrian rocks (short-dashed lines). Ore bodies and their nomenclature after Radabaugh, Merchant, and Brown (1968).

consists of vertically elongate replacement chimneys in dolomite rocks. The main chimneys are at or near the lower ends of the mantos (fig. 11). The chimneys extend from the top of the Leadville Dolomite through the Gilman Sandstone and Dyer Dolomite to bases in the Parting Formation or Harding Sandstone, and, in general, they taper downward. The chimneys are composite ore bodies consisting of two quite different suites of sulfide minerals. The older and more extensive of these suites consists in part of pyrite, which forms massive cores of the chimneys, and in part of sphalerite, which is concentrated in shells surrounding the central pyrite bodies. These shells merge with the mantos and are nearly surrounded by an outer shell of siderite. The younger suite consists of copper, silver, lead, and gold minerals that irregularly encrust and replace parts of the pyritic cores of the chimneys. Ore containing these minerals — some of it very rich — is called copper-silver ore by the mine staff and is also known colloquially as "chimney ore."

A third group of ore bodies, far subordinate in value of output to those just discussed, consists of pyritic deposits in the Rocky Point zone of the Sawatch Quartzite. Many of the ore bodies of this group are beneath or on the trend of the mantos in the Leadville Dolomite, but some, notably the Bleak House, Pine Martin, and Ground Hog mines, are in flanking or intermediate positions (fig. 11). The deposits in the Rocky Point zone have been worked principally in the zones of oxidation and secondary sulfide enrichment, and the principal output has been in gold and silver. Mineralization in the Rocky Point zone was in two main stages. In the first stage, pyrite was deposited widely as a dissemination in certain beds of quartzite and as a filling or coating along fractures and solution cavities. In the second stage, siderite, pyrite, and local marcasite, accompanied by small amounts of sphalerite, chalcopyrite, galena, barite, gold, and silver, were deposited in fractures and openings within sharply defined anastomosing swaths within either the pyritized quartzite or previously unmineralized quartzite. Oxidized ore was mined principally along these swaths or channelways.

Silver-and-gold telluride veinlets and pockets in the Sawatch Quartzite constitute a fourth type of deposit. Tiny telluride veinlets occur on random joint surfaces in the lower 30–35 ft (9–10.7 m) of the quartzite in the Ben Butler, Tip Top, and Star of the West mines near Belden (fig. 11), where they were encountered and were mined on a small scale in the course of mining along veins in the underlying Precambrian rocks. The veinlets consist principally of hessite but contain some petzite and a little native gold. Pockets of telluride ore were also encountered in early-day mining in the Rocky Point zone of the Sawatch Quartzite. The telluride minerals in scattered veinlets and pockets probably were a source of some of the silver and gold in the oxidized ores in the Rocky Point zone.

Veins in the Precambrian rocks have been explored rather extensively but have been the source of only a very small fraction of the output from the Gilman district. The most productive veins — in the Star of the West, Ben Butler, Tip Top, and Mabel mines — are in a cluster that projects beneath No. 3 manto in the Leadville Dolomite (fig. 11), though no physical connection has been found between the veins and the manto. The veins have been mined principally for their gold and silver content. They consist mainly of quartz and pyrite but locally contain abundant chalcopyrite, galena, and sphalerite in small ore shoots. The mineralized streaks in most of the veins are less than 2 ft (60 cm) wide, though the fault zones occupied by the streaks may be much wider. The largest and most productive vein — the Mabel — locally reached a width of 8 ft (2.4 m) (Crawford and Gibson, 1925, p. 70).

A sixth and last type of deposit occurs in the Leadville Dolomite in several old and small mines at Red Cliff (pl. 1). These mines were inaccessible to us, and we rely on the descriptions of Crawford and Gibson (1925) for information concerning them. The small ore bodies in these mines were characterized by a high silver content in a pyrite-galena-sphalerite ore or its oxidized equivalent. In this respect, the deposits suggest an alliance with the chimney deposits at Gilman. The deposits occur in strongly fractured dolomite rock, both as small veins of limited extent and as replacement bodies in dolomite. This combination of fracture and replacement occurrence is similar to that of the Rocky Point zone in the Sawatch Quartzite. Small bodies of jasperoid occur with or in the vicinity of the deposits near Red Cliff. In this respect, the deposits differ from the mantos and chimneys at Gilman, which have no jasperoid closely associated with them. The Red Cliff deposits seem to constitute a small mineralized center distinct from that at Gilman.

MINE WORKINGS

The upper ends of the manto ore bodies were worked originally from their outcrops on the wall of Eagle Canyon through several separate mines, notably the Little Chief, Iron Mask, Belden, and Black Iron mines (fig. 11). These mines were subsequently consolidated and extended to form the present Eagle mine, which encompasses practically all the workings in dolomite rocks. At an early stage in the consolidation, the now-obsolete terms "Eagle No. 1" and "Eagle No. 2" were applied, respectively, to the northwestern (Little Chief and Iron

Mask) and the southeastern (Black Iron) workings (Crawford and Gibson, 1925). Chief access to the Eagle mine is through the Wilkesbarre shaft at Gilman, and through the Newhouse tunnel, the main ore haulageway (fig. 11). The Wilkesbarre shaft, 400 ft (122 m) deep, extends to 16 level, the arterial level of the mine. The Newhouse tunnel, the portal of which is near the railroad at Belden, is continuous with the 16 level workings.

The key workings of the mine are long strike drifts on levels 14 through 20. These drifts are in the lower part of the Leadville Dolomite or in the Dyer Dolomite and, hence, in the footwall of the ore bodies in the upper part of the Leadville. Access to the levels below 16 is through 2 inclines with slopes of about 12°. Inclines and raises connect the workings on 14 and 15 levels to 16 level. Workings above 14 level are those of the old mines that were worked through inclines from the outcrop. At the time of our studies, 20 level was the bottom level in the mine. In a major exploration effort in later years, inclines were driven from 20 level as deep as 28 level, and extensive drifts were driven at several intermediate levels. Several small bodies of ore were found, but none of the magnitude of the mantos and chimneys shown in figure 11. In all, the workings of the Eagle mine, totaling more than 65 mi (105 km), extend 8,500 ft (2.6 km) downdip from the outcrops of the ore bodies and 9,000 ft (2.75 km) along the strike of the beds (Radabaugh and others, 1968). The deepest workings are about 2,800 ft (854 m) below the upper slopes of Battle Mountain. Though 18 level is at about the level of the Eagle River, only small amounts of water were encountered in deeper workings; this is consistent with the absence of through-going fractures from the Precambrian into the sedimentary rocks.

All the ore mined is raised or lowered to 16 level and is hauled via the Newhouse tunnel to an underground 1,200-ton differential flotation mill just off the tunnel near its portal. The underground location of the mill was dictated by the absence of any suitable space on the sides or bottom of the steep canyon of the Eagle River. The mill chamber, 52 ft (16 m) wide, 332 ft (101 m) long, and 22–56 ft (7–17 m) high, is in Cross Creek Granite. The mill and chamber were described by Borcherdt (1931).

All the workings of the mines in the Rocky Point zone of the Sawatch Quartzite are in the inclined plane of the bedding. The main workings are inclines (inclined adits) from which branch a maze of inclined lateral workings and stopes of very low ceilings (fig. 50; pl. 2*A–D*). Many of the inclines open on a cliff face. With the rotting away of access structures, the inclines are virtually inaccessible, though they remain open.

PRODUCTION AND MINING HISTORY

An exact record of production from the Gilman district does not exist because the production statistics of the U.S. Bureau of Mines are kept by county rather than by district. The record for Eagle County, summarized for 5-year intervals from 1880 through 1972, is shown in table 2. At least 99 percent of the production total of more than $329 million shown in the table came from the Gilman district. The remainder came from several minor sources: (1) the Taylor Hill area (Emmons, 1886, p. 534; Tweto, 1956), 10 mi (16 km) southeast of Red Cliff, which possibly produced $250,000 in gold, mainly before 1890; (2) the Holy Cross City district (Tweto, 1974), 10 mi (16 km) southwest of

TABLE 2.—*Production of gold, silver, copper, lead, and zinc in Eagle County, Colo., 1880 through 1972,*

[Data for period 1880 through 1907 compiled from unpublished sources by C. W. Henderson, U.S. Bureau of Mines. Data for period 1908–72 from U.S.

Years	Gold		Silver		Copper	
	Ounces (Troy)	Value	Ounces (Troy)	Value	Short tons	Value
1880–84	5,500	$111,000	603,414	$675,884	...	...
85–89	44,454	910,333	1,357,911	1,337,202	...	...
90–94	28,329	586,002	953,588	844,889	...	...
95–99	7,671	158,804	280,467	175,510	41	$10,300
1900–04	13,477	279,045	509,593	305,106	366	112,941
05–09	12,727	263,532	423,914	250,554	263	78,591
10–14	9,861	204,099	796,617	462,233	248	69,373
15–19	13,930	288,559	849,264	670,484	352	162,859
20–24	11,314	234,279	2,078,594	1,976,314	2,335	651,055
25–29	5,745	118,868	1,455,059	815,175	2,284	756,376
30–34	23,576	608,302	7,654,961	3,146,408	18,049	3,027,259
35–39	74,816	2,618,562	21,944,632	15,575,490	47,945	9,682,828
40–44	59,290	2,073,750	11,824,982	8,408,876	16,654	3,831,204
45–49	5,281	184,835	972,507	865,053	682	281,965
50–54	24,000	840,000	4,123,225	3,731,727	3,594	2,006,558
55–59	20,397	713,895	4,807,464	4,350,997	6,428	4,253,683
60–64	24,565	859,775	3,472,611	3,824,999	3,845	2,411,746
65–69	6,378	231,191	1,550,186	2,357,604	1,638	1,224,737
70–72	2,130	95,061	411,555	682,979	77	81,979
Total	393,491	11,379,892	66,070,544	50,457,484	104,801	28,643,454

Gilman, which may have produced as much as $300,000 in gold, mainly from 1880 to 1910; (3) the Fulford area (Gabelman, 1950), 15 mi (24 km) west of Gilman, which produced a small amount of gold — probably less than $100,000; (4) the Brush Creek area southeast of Eagle, which has a recorded production of 200,000 ounces of silver (worth about $125,000) in the period 1913–18 (Henderson, 1926, p. 47); and (5) the State Bridge area (fig. 1) which has produced very small amounts of copper and placer gold. At the most, total production from these minor districts has not exceeded $1 million. Thus, the value of production from the Gilman district through 1972 is about $328 million as valued at the times of sale.

The first discoveries of ore deposits in the Gilman–Red Cliff area were made in 1879 during the wave of prospecting that followed the discoveries at Leadville in 1876. Many of the main deposits of oxidized silver-lead ore in the Leadville Dolomite at Gilman and at Red Cliff were located in that year, as were also some of the oxidized gold-silver deposits in the Sawatch Quartzite. Among the mines that were in production by 1880 were the Belden, Black Iron, and Little Chief mines near Gilman (fig. 11) and the Horn Silver and Wyoming Valley mines near Red Cliff (pl. 1). Upon inception of mining, the town of Red Cliff was established, and the Battle Mountain Mining District was organized in what was then a part of Summit County. A smelter was completed at Red Cliff in 1880. In November 1881 the narrow-gage Denver and Rio Grande Railroad reached Red Cliff from Leadville and was continued through the canyon of the Eagle River as far as Rock Creek (pl. 1), which remained the terminus until 1887. To transport ores from mines located high on the cliffs of the canyon, aerial or rail trams were constructed from many of the mines down to the railroad; remnants of these structures still cling to the canyon walls.

In 1883 the district reached a production rate of 8,000 short tons of lead and 232,000 ounces of silver, and largely on the strength of its prosperity, Eagle County was organized from a part of Summit County, with Red Cliff as the county seat.

Intensive mining of the oxidized gold-silver ores in the Rocky Point zone of the Sawatch Quartzite began in 1884 and is reflected in an early peak in gold output through the period 1885–94 (table 2). During this period, production from the mines in the Leadville Dolomite began to decline as the workings passed downward from the rich lead-silver ores of the oxidized zone into sulfide ores. The sulfide ores first encountered consisted principally of zinc, for which no market nor treatment facility existed at that time in Colorado.

Extraction of zinc began in 1905 with the construction of a roasting and magnetic separation plant by the Pittsburg Gold-Zinc Co. and by its successor, the Eagle Mining and Milling Co. The iron oxide waste from this plant was conveyed by aerial tramway up onto the west wall of the canyon, where it forms a prominent bare rusty landmark immediately south of Fall Creek.

In 1912 the New Jersey Zinc Co. began to acquire holdings in the vicinity of Gilman through its subsidiary, the Empire Zinc Co. By 1918 this company had acquired the Iron Mask group of the Eagle Mining and Milling Co. and the Black Iron group of the American Zinc Co., as well as other properties, and the company consolidated them into a single operating unit called the Eagle mine (fig. 11), which was still in operation in 1974. Zinc ore produced on a large scale from these workings in the period 1915–30 was treated in a roasting and magnetic separation plant at Canon City, Colo. Silver, gold and copper produced in the district during this period (table 2) came in part from the mines at Red Cliff, and in part from the Ground Hog, Percy Chester, and Rocky Point mines in the Sawatch Quartzite and the Mabel and Ben Butler mines in the Precambrian rocks.

In the late 1920's the group of chimney ore bodies near the lower ends of the mantos in the Eagle mine (fig. 11) was reached. Discovery of these ore bodies, together with declining prices of zinc, led to a halt in the mining of zinc ore in 1931. Production of zinc ore was not resumed until 1941. The chimney ores account for the large production of silver, copper, and gold in the period 1930–45 and for significant though declining amounts of these metals thereafter (table 2). As a consequence of the chimney ores, the Gilman district leads all other districts in the State in the production of copper.

Since 1941 zinc has been the principal product of the

in terms of recoverable metals

Bureau of Mines, Denver, Colo. Leaders indicate no production]

Lead		Zinc		Total value
Short tons	Value	Short tons	Value	
13,200	$1,147,000	...	...	$1,933,884
6,772	560,794	...	...	2,808,329
8,518	668,749	...	...	2,099,640
3,082	227,967	...	...	572,581
4,171	359,996	10	$880	1,057,968
411	42,193	1,801	188,024	822,894
2,512	217,179	14,556	1,662,952	2,615,836
4,322	606,803	40,700	9,208,005	10,936,710
1,683	256,305	31,693	4,209,292	7,327,245
6,331	935,008	20,888	3,023,265	5,648,692
6,917	582,095	27,525	2,377,796	9,741,860
3,295	318,677	...	...	28,195,557
8,569	1,131,747	83,107	16,792,351	32,237,928
4,905	1,421,246	83,422	20,528,436	23,281,535
15,452	4,694,176	110,610	32,822,268	44,094,729
18,218	4,977,741	113,246	26,658,534	40,954,850
19,151	4,206,477	119,405	29,067,150	40,370,147
11,748	3,462,190	121,748	34,644,974	41,920,696
9,187	2,700,658	89,151	29,055,401	32,616,078
148,444	28,517,001	857,862	210,239,328	329,237,159

Eagle mine and the Gilman district. About 1970 the district surpassed the Leadville district in total zinc output and became the State's leading zinc district in terms of total as well as annual output.

In addition to base and precious metals, the Gilman district has produced about 200,000 tons of manganiferous iron ore (Umpleby, 1917). Part of this total was supplied to the steel plant at Pueblo, Colo., in the 1890's, and part was shipped to smelters as a fluxing ore in the early 1900's. No production has been made since that time, though Umpleby (1917) estimated that 750,000 tons of material containing 12–15 percent manganese and 38 percent iron remained in 1917. This reserve is principally in the old Black Iron and Iron Mask mines (fig. 11) where it formed extensive parts of the ore bodies in the oxidized zone. Both mines took their names from the black mixtures of iron and manganese oxides that marked the outcrop of the ore bodies in Leadville Dolomite.

Details of the early history of the Gilman district may be found in Henderson (1926), Crawford and Gibson (1925) and, especially, in the annual reports of the Director of the Mint for the years 1880–96.

ROCK ALTERATION

Many varieties of altered rock are associated with the ore deposits of the Gilman district. This assortment reflects two factors: (1) the differences in composition of materials as varied as carbonate rocks, quartzite, and granite, and (2) changes in the character of alteration with time. Most of the alteration is specifically related to the ore deposits in space and time, but some — notably the dolomitization of the Leadville Limestone — is related to processes that affected a large area of the mineral belt, far beyond the confines of the Gilman district. Though much of the alteration preceded mineralization (deposition of ore and gangue minerals), the later stages of alteration evidently proceeded concurrently with mineralization. We refer to the entire sequence of alteration and mineralization stages as the hydrothermal stage. In this application, we use "hydrothermal" in the literal sense of warm (or hot) water, without implication as to the source of the water or of its dissolved chemicals.

In the discussion that follows, we shall consider — in the order named — the alteration facies or stages in the several rock units affected, which are (1) the carbonate rocks (Leadville and Dyer Dolomites), (2) karst silt, (3) Gilman Sandstone, (4) Sawatch Quartzite, (5) Precambrian rocks, and (6) porphyry sill. Some of these — notably the carbonate rocks — exhibit several alteration facies, which we shall examine in age sequence from oldest to youngest. Discussion of the relationships between alteration stages in the various rock settings and of alteration stages to mineralization stages will be found in the section "Summary of Hydrothermal History."

CARBONATE ROCKS

Alteration in the carbonate rocks began with the transformation of the limestone of the Leadville Limestone to dolomite. This process occurred throughout the 40-mile (65-km) width of the mineral belt and thus was not unique to the Gilman district. We refer to the product of this dolomitization of the Leadville as the early dolomite. The Dyer Dolomite was not affected; the rock of this unit is inferred to have been a dolomite since deposition in Devonian time or shortly thereafter.

The early dolomite of the Leadville was later partly and irregularly recrystallized. Again, the recrystallization was not unique to the Gilman district but occurred in many areas, some of which were later mineralized and some of which were not. Two common forms of the recrystallized dolomite are (1) zebra rock, a banded or thinly layered rock consisting of generally fine grained dark dolomite and coarse grained white dolomite in alternating thin layers, and (2) pearly dolomite, a medium- to coarse-grained light-gray to light-brownish-gray dolomite with a pearly luster. Some pearly dolomite formed by recrystallization of zebra rock, but, mainly, the pearly dolomite is thought to have recrystallized directly from the early dolomite of darker color and finer grain. The Dyer Dolomite was affected only to a very minor degree by these changes; small patches of zebra rock occur in it only in widely scattered localities. The distinctive dolomite of the pink breccia — at a fixed stratigraphic level in the Leadville Dolomite — crystallized after zebra rock, probably during formation of some of the pearly dolomite.

After this series of changes in the Leadville, processes occurred that affected both the Leadville and the Dyer. These processes, which may have operated concurrently in different localities, were (1) dissolution of dolomite, starting with "sanding" and leading ultimately to the formation of openings, such as channelways and caves; (2) replacement of dolomite by jasperoid in certain localities; and (3) the concentration of residues from the dissolution or replacement of dolomite as greasy black clays.

The various types of alteration in dolomite are discussed in detail in the following sections.

EARLY DOLOMITE

The early dolomite of the Leadville resulted from the complete dolomitization of a nearly pure limestone. A sample of the limestone from a locality near Minturn,

1.5 mi (2.5 km) north of the Gilman district, consists petrographically of calcitic foraminiferal fragments in a matrix of very finely granular calcite and small amounts of iron-stained interstitial matter. An analysis (table 3, No. 1) shows 1.28 percent insoluble matter, which consists of slender needles of authigenic quartz and a small amount of black bituminous or carbonaceous organic matter. The soluble fraction has a calculated composition of about 97 percent calcite and 3 percent dolomite and ferrodolomite. At localities more distant from Gilman the limestone is even more pure. Engel, Clayton, and Epstein (1958, p. 376–77) reported that the upper part of the Leadville consists of 99.0 percent or more calcite, and that, exclusive of chert, the lower part contains 1.5–3.15 weight percent silica in disseminated microcrystalline form and less than 0.5 weight percent of other elements, mainly Al, Fe, and Mg. Engel and Engel (1956) reported that on the basis of 75 analyses from 25 localities, the Cu, Pb, and Zn contents of the limestone vary little from 3, 3, and 10 ppm (parts per million), respectively.

The predominant type of early dolomite is dark gray to black and finely crystalline. As seen in thin section it consists principally of interlocking dolomite grains 0.02 to 0.04 mm in diameter but contains coarser grains in scattered lenses as much as 1 mm long. The principal visible impurity is organic matter — the cause of the dark color — which is evenly dispersed through the dolomite grains in dustlike particles less than 0.004 mm in diameter and also occurs as scattered black intergranular specks as much as 0.05 mm in diameter. The only other impurity noted is in the form of minute clear inclusions in some of the dolomite grains. The inclusions have a refractive index of less than 1.50 and are evidently fluid.

A subordinate type of the early dolomite is lighter in color (medium gray) and coarser in grain size than that just described. It occurs in scattered beds in the part of the Leadville below the pink breccia and is abundant above the pink breccia. In this rock the grain size averages perhaps 0.1 mm, and the organic pigment is largely concentrated at grain boundaries.

On the basis of chemical analysis (table 3, No. 2), the dark fine-grained dolomite has a calculated composition of about 98 percent dolomite, 1.5 percent ferrodolomite, and 0.5 percent magnesite molecules. The insoluble residue consists entirely of a fine dust of organic matter. An old analysis (table 3, No. 6) suggests that the coarser and lighter type of the early dolomite contains nearly 2 percent of the magnesite molecule.

Porosity and permeability of the early dolomite are low. Roach (1960) measured porosities of only 2–6 percent in samples not affected by sanding. Rove (1947, p. 72) and Wehrenberg and Silverman (1965, p. 330) measured permeabilities on the order of 10^{-4} to 10^{-6} millidarcys in the early dolomite — lower than in the undolomitized Leadville Limestone (10^{-4} to 10^{-5} millidarcys).

Thus, if volume changes occurred as a result of dolomitization, they are not expressed by increase in porosity. They could be expressed by a shrinkage in thickness of the Leadville, but this possibility cannot be appraised because of the unevenness of the erosion surface at the top of the Leadville, because of the effects of both karst and hydrothermal dissolution processes, and because of the absence of suitable exposures at the dolomite-limestone interface.

The processes, causes, and timing of the dolomitization of the Leadville are discussed later. We note here, however, that, by our interpretation, the early

TABLE 3. — *Composition of limestone and dolomite, Gilman area*

[Analyses 1–5 by V. North, U.S. Geological Survey. CO_2 calculated by difference. Values in percent]

Sample No.	Chemical analyses						Calculated molecular composition of carbonate			
	CaO	MgO	FeO	MnO	CO_2	Insoluble	Calcite $CaCO_3$	Dolomite $CaMg(CO_3)_2$	Ferrodolomite[1] $CaFe(CO_3)_2$	Magnesite $MgCO_3$
1.	54.72	0.41	0.34	0.010	43.24	1.28	96.98	2.01	1.01	0.00
2.	30.50	21.81	.57	.016	46.69	.42	.00	98.08	1.46	.45
3.	30.48	21.71	.50	.014	46.91	.39	.00	98.53	1.29	.18
4.	30.34	21.87	.36	.017	47.13	.28	.00	98.62	.92	.46
5.	30.79	21.11	.61	.023	47.04	.43	1.48	96.86	1.66	.00
6.	29.71	21.95	.55	.00	46.84	.24	.00	96.76	1.30	1.94

[1] The small amounts of MnO present are calculated as ferrodolomite.

SAMPLE DESCRIPTIONS

1. Foraminiferal limestone in Leadville Limestone about 100 ft (30 m) above base. Contains a little authigenic quartz and rusty interstitial matter. Locality 3,000 ft (914 m) S. 64° W. of Minturn roundhouse (Tweto and Lovering, 1977, pl. 1).
2. Fine-grained black dolomite about 50 ft (15 m) above base of Leadville. Knob on south side of Bolts Lake, 2 mi (3.2 km) northwest of Gilman (Tweto and Lovering, 1977, pl. 1).
3. Fine-grained dark dolomite in zebra rock, about 70 ft (21 m) above base of Leadville. Crosscut northeast from Wilkesbarre shaft, 16 level. Eagle mine.
4. Coarse-grained gray vuggy dolomite that replaces dolomite of sample No. 2. Contains a very little quartz and pyrite.
5. Coarse-grained brown-gray dolomite that replaces the zebra rock of sample No. 3.
6. Fine-grained dolomite 35 ft (10.6 m) below top of Leadville in cliffs in gulch just south of Gilman; P. M. Dean, analyst (Crawford and Gibson, 1925, p. 60).

dolomitization resulted from quite different causes and processes than many of the later types of alteration and attendant mineralization and that the two events were separated in time by at least several million years.

ZEBRA ROCK AND PEARLY DOLOMITE

Zebra rock and pearly dolomite occur in scattered bodies or beds throughout the Leadville Dolomite but are most abundant in the upper part of the formation. At Gilman, this is the part above the pink breccia, about 90 ft (27 m) above the base of the Leadville. Where the preserved thickness of the Leadville is less than 90 ft, these two varieties of dolomite rock occur at stratigraphic levels that at Gilman are mainly early dolomite. Two miles south of Red Cliff, for example, where only the lower 67 ft (20.4 m) of the Leadville is preserved beneath the pre-Belden karst erosion surface, zebra rock and pearly dolomite constitute about 30 percent of the formation (Tweto, 1949, p. 185). At Leadville, where the formation locally thins to zero, these two rock types in many places constitute a major part of whatever thickness is preserved (Tweto, 1968a).

Zebra rock, a product of partial recrystallization of the early dolomite, is characterized by alternating thin layers of dark and light dolomite (figs. 12, 13). The individual layers in typical zebra rock are 1–10 mm thick, pinch and swell irregularly, and generally lens out in short distances. The layering in most zebra rock is approximately parallel to the bedding, but in places it is at high angles. The layering in some patches of zebra rock ends abruptly against joints, suggesting that the joints existed before the zebra rock formed. In places regular layering fades into a disorganized pattern of patches of white dolomite scattered in fractured dark dolomite, as shown in the lower part of figure 12. Many varieties of zebra rock exist, depending on the proportions of light and dark dolomite, the relative grain size of the two components, and the degree of recrystallization of either the dark component or of both components.

The dark layers in the simplest and most abundant variety of zebra rock are unaltered remnants of the early dolomite. (Compare analyses 2 and 3, table 3.) The dolomite of the light layers is coarser than that of the dark (fig. 14). This dolomite has the same refractive index as that of the dark layers (n_O, 1.681–1.682), and, as first noted by Emmons (1886, p. 64), it is identical chemically. However, the two varieties differ isotopically, as discussed under "Properties of Dolomite Rocks." The only visible difference between the dark and light layers is in grain size and in the much greater abundance of pigmenting organic material in the dark layers. Black organic matter is commonly concentrated at the boundary between the dark and light layers, as if expelled from the recrystallized light layers. In other varieties of zebra rock, the dark layers are recrystallized to a form that is coarser in grain size and paler than the original early dolomite. The paler color reflects a reduction in the content of pigmenting organic matter. Evidently, some organic matter was lost in the recrystallization process. In a still more advanced stage of recrystallization, the dark and light layers of zebra rock were destroyed and were replaced by homogeneous light-gray or tan coarse-grained pearly dolomite. This dolomite differs from its predecessors in containing a slight excess of Ca rather

FIGURE 12. — Zebra rock in wall of drift, 16 level, Eagle mine. Area of photograph about 3 × 6 ft (1 × 2 m).

FIGURE 13.—Specimens of zebra rock. Lower specimen displays advanced stage of recrystallization to pearly dolomite. Specimens are about 3 in. (7.6 cm) long. After Behre, 1953, p. 36.

than Mg (table 3, No. 5). The recrystallization was accompanied by a further loss of organic matter, leaving only trace amounts scattered uniformly through the rock.

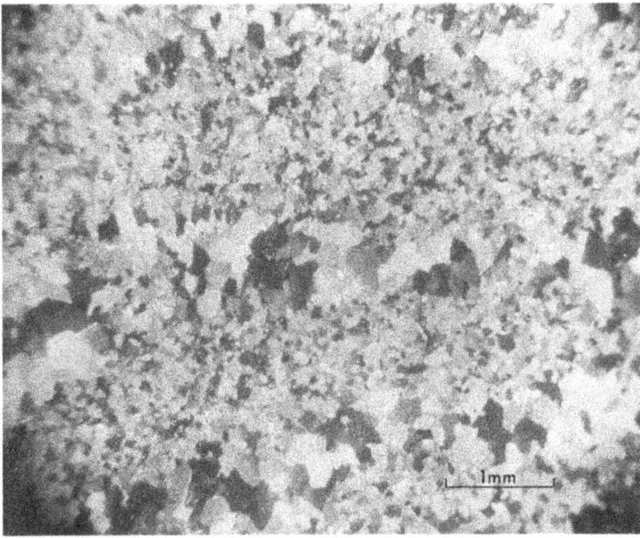

FIGURE 14.—Photomicrograph of zebra rock, showing contrast in grain size between white layers (belts of coarse crystals) and dark layers (belts of fine crystals). Crossed nicols.

Pearly dolomite that recrystallized directly from early dolomite does not differ visibly from that just described, but an analysis (table 3, No. 4) does not show excess Ca.

Zebra rock and pearly dolomite vary widely in degree of porosity. In some places, they have no visible porosity and seem as dense as the early dolomite; in others, they have porosity that is readily visible with a hand lens; in still others, they contain openings visible to the naked eye. Systematic measurements of the porosity have not been made. On the basis of a few measurements, permeabilities range from 2.85 to 10^{-3} millidarcys (Wehrenberg and Silverman, 1965, p. 330), considerably higher than for the early dolomite.

The similarity of zebra rock and pearly dolomite to recrystallized dolomites in many other areas is worthy of note. Zebra rock like that of the Gilman district and the mineral belt occurs in the Metaline zinc-lead district of Washington (Park and Cannon, 1943, p. 42–43; Dings and Whitebread, 1965, p. 15). The "coontail ore" of some mining districts of the Eastern United States — illustrated by Grogan and Bradbury (1968, fig. 7) — has the fine-layered structure of zebra rock, though different in composition. The so-called recrystalline dolomite of the Mississippi Valley and East Tennessee mining areas closely resembles the pearly dolomite and also contains material similar to zebra rock that is in an advanced stage of recrystallization. (For example, see Ohle, 1951; Kendall, 1960; Hoagland and others, 1965.) Whatever the causes of dolomitization and the recrystallization of dolomite, the Gilman district and the Colorado mineral belt share the riddle with many other localities.

FIGURE 15. — Pink breccia bed of the Leadville Dolomite in irregular contact with overlying dark dolomite (above hammerhead and at right), 16 level, Eagle mine.

PINK BRECCIA

The pink breccia is a bed of the Leadville Dolomite that is characterized by distinctive lithology. As such, it is an important stratigraphic marker readily identified in mine workings and in drill cores or cuttings. The bed was originally about 2 ft (0.6 m) thick, but in consequence of bedding-fault movements and dissolution, it now pinches and swells through a range of 1–4 ft (0.3–1.2 m) in most places. The pink breccia was interpreted as an intraformational breccia of sedimentary origin by geologists of the New Jersey Zinc Co. (Radabaugh and others, 1968, p. 648). We do not necessarily dispute this interpretation, but we believe it more significant that (1) the bed marks a minor unconformity within the Leadville and contains a larger proportion of detrital material and iron than other parts of the Leadville; (2) brecciation was, at the least, greatly increased by bedding-fault movements; and (3) dissolution processes contributed further to the breccia structure and also produced a residual black clay that in many places is a prominent component of the breccia.

Typical pink breccia (fig. 15) consists of ellipsoidal or augenlike pieces of light-pinkish-gray dolomite of various sizes in a matrix of black clay, loosely granular dolomite, or green-gray shaly detrital material. In some places the dolomite clasts are angular, and in a few places the dolomite of the bed is a simple crackle breccia. Both the upper and lower contacts of the breccia are generally wavy, and locally the breccia is markedly discordant with the bedding (fig. 16). The discordant relations evidently reflect late bedding-fault movements, some of which caused plastic flow like that of an intrusion breccia.

The dolomite of the pink breccia typically has a sugary texture, but in some places it is dense, breaks conchoidally, and superficially resembles white chert. As seen in thin section, the sugary dolomite consists principally of interlocking grains 0.2–0.5 mm in diameter; it also contains a few veinlets of slightly coarser dolomite and scattered globular bodies 3–4 mm in diameter of coarsely crystalline white dolomite. Interstitial quartz or chalcedony is locally abundant. The dolomite grains contain small amounts of opaque black dustlike material, and in pink dolomite, they contain abundant minute particles of hematite. Hematite also

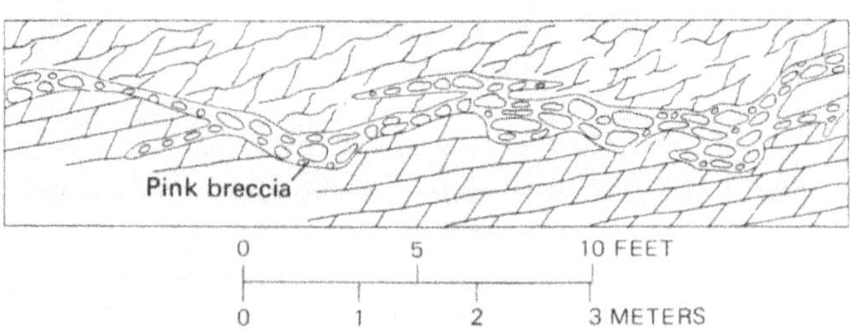

FIGURE 16. — Sketch of discordant relations of pink breccia, 16 level, Eagle mine. Rock below the pink breccia is evenly bedded dark dense (early) dolomite. Rock above the pink breccia is crackle breccia of zebra rock and pearly dolomite.

occurs locally as intergranular hexagonal plates. Scattered grains of pyrite less than 0.1 mm in diameter are present in some places, especially near ore. Detrital grains of quartz, tourmaline, and zircon are present in some samples and absent in others.

The dolomite of the pink breccia is a recrystallization product of early dark dolomite. In places, tongues of the pink dolomite showing replacement relations extend into dark dolomite above or below the pink breccia. Elsewhere, the part of blocks of dark dolomite that project into the breccia have rinds of pink to white recrystallized or decolorized dolomite. Similarly, some of the larger ellipsoidal fragments in the pink breccia have cores of dark dolomite and outer shells of pink to white dolomite (fig. 15). In places the breccia contains pieces of zebra rock, some of which are largely replaced by pink dolomite, indicating that some of the pink dolomite postdates zebra rock. In some localities where pink breccia is discordant and, to a degree, intrusive, it cuts zebra rock (fig. 16).

As the dolomite of the pink breccia is the only variety of dolomite in the Leadville that contains hematite, we infer that the parent rock of this bed was richer in iron than the remainder of the Leadville. This would not be inconsistent with the unconformity recognized at this horizon nor with the presence of green, presumably iron bearing, shaly material in parts of the pink breccia. The black clay in the pink breccia is a product of dissolution after crystallization of the pink dolomite. It is discussed separately, along with other occurrences, in a following section.

SANDING AND CHANNELING

At a stage following those just discussed, the dolomite of the Gilman district and of many other areas in the mineral belt was exposed to solutions in which it was markedly unstable. Dissolution at this stage affected all the varieties of dolomite in the Leadville and, also (though less widely) the syngenetic or diagenetic dolomite of the Dyer. The dissolution began before the deposition of sulfide minerals but continued during the sulfide stage.

Dissolution began with the disaggregation, or sanding, of dolomite. Products of this process range from slightly friable dolomite to free-running dolomite sand. Thin sections of friable dolomite show that dissolution proceeded along intergranular boundaries, changing the rock from a complex of interlocking crystals to a porous aggregate of subrounded grains. The end product in free-running sand is composed of round to subround dolomite grains, each of which is a remnant of one, or parts of only a few, former crystals.

By the time of sanding, the dolomite rocks had been sliced or diced by countless joints. The sanding solutions clearly gained access to the otherwise dense and nearly impermeable dolomite through the joints. In areas of incipient dissolution, the sanding is confined to narrow sheaths enclosing joints. Very commonly, only the joints of one direction are affected. In areas of more intense sanding, blocks outlined by joints are observed to have been attacked from all sides.

Areas affected by sanding are more abundant near ore bodies than elsewhere, but they are by no means restricted to the vicinity of ore. At the northwest end of 18 level of the Eagle mine, 1800 ft (550 m) from the nearest known ore body, many dozens of mine-carloads of dolomite sand flowed into the drift through an opening only a few feet (1–2 m) in diameter. This sand came from the Leadville above the pink breccia. At many places on 14 and 15 levels, large pockets of dolomite sand were encountered in the Dyer, far below the ore bodies near the top of the overlying Leadville. Many areas of sanded dolomite occur in the absence of known mineral deposits in the area between Gilman and Leadville.

Continuation of dissolution beyond the sanding level of intensity produced open channelways and caves. These openings are most abundant near ore but are also found in many other places. Many of the openings were partly or wholly filled with detritus of two kinds — resedimented dolomite sand and cave rubble. The resedimented dolomite sand (fig. 17) is generally finely stratified and in places shows features such as crossbedding, ripple marks, scour-and-fill structures, and various penecontemporaneous compaction and slump structures. The sand probably was transported by the same fluids that created the openings. The sedimentary structures found in the sand suggest that, in places at least, these fluids flowed briskly.

As channels and caves grew, rockfalls occurred, forming cave rubble (figs. 17, 18). By continuing roof collapse, some caves extended from positions deep in the Leadville, or even in the Dyer, up to the Belden Formation and the porphyry sill (Pando Porphyry) within it. Some also interesected the silt-filled preexisting channels of the old karst channel system near the top of the Leadville. Thus, fragments of shale and sandstone from the Belden, chert-rich karst silt, and porphyry from the sill occur along with dolomite fragments and sand in caves or channels as deep as the Dyer.

Dissolution both preceded and accompanied deposition of sideritic sulfide ore. Cave sand and rubble were extensively replaced by ore (figs. 18, 42) and therefore predate the ore. However, there is also good evidence that sanding and channeling accompanied ore deposi-

FIGURE 17. — Stratified dolomite sand resting on irregular floor of cave (lower right) and overlain by cave rubble (top of photo). "Run" of loose sand at left, 17 level, Eagle mine. Area of photo is about 5 × 10 ft (1.5 × 3 m).

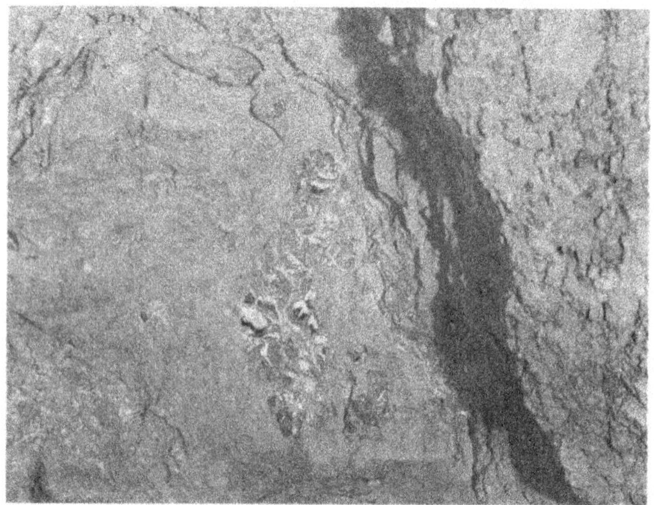

FIGURE 18. — Corroded rockfall blocks of sanded zebra rock enclosed in faintly stratified and crenulated dolomite sand. Sulfide ore at right has replaced cave rubble. Steep dark streak is black clay selvage of sulfide body. Area of photograph is about 2 × 4 ft (0.6 × 1.2 m).

tion. Fragments of sulfide ore and siderite are present in some unmineralized cave rubbles. In at least one locality (fig. 19), a cave rubble contains fragments of the paragenetically old minerals pyrite and siderite but is encrusted by paragenetically younger sphalerite. Thus, the rubble evidently formed between the pyrite-siderite and the sphalerite stages of deposition at this particular locality. Time relations between dissolution and stages of ore deposition are discussed further under "Summary of Hydrothermal History."

PROPERTIES OF DOLOMITE ROCKS

Many investigations of the chemical, physical, and isotopic properties of the Leadville and Dyer Dolomites have been made either for application to exploration or for their bearing on mineralization processes. For the purposes of many of these studies, distance from ore rather than the kind of dolomite was the paramount concern.

Engel and Engel (1956) made 50 analyses of individual dolomite grains from ore contacts and the walls of channels and found that the Cu, Pb, and Zn contents varied little from 3, 3, and 10 ppm, respectively, the same as for unaltered Leadville Limestone. Thus, as a *mineral*, dolomite seemingly registers no change in heavy-metal content as a result of exposure to mineralizing solutions. Wehrenberg and Silverman (1965, p. 333) investigated the metal content of dolomite *rock*, which in addition to dolomite mineral grains would contain any filling added in pore spaces and in microfractures by mineralizing fluids. They found Pb and Zn contents as high as thousands of parts per million in sanded dolomite adjacent to ore in several localities and a fairly systematic decrease in metal content outward to distances of 64 ft (19.5 m) from ore bodies. Unsanded dolomite at distances of 2–4 ft (0.6–1.2 m) from ore showed metal contents one or two orders of magnitude lower than in sanded dolomite.

Isotopic analyses of the lead in dolomite and limestone (Engel and Patterson, 1957) show that the lead (3 ppm) in dolomite near the northwest edge of the dolomitized belt (northwest of Gilman) is of the same

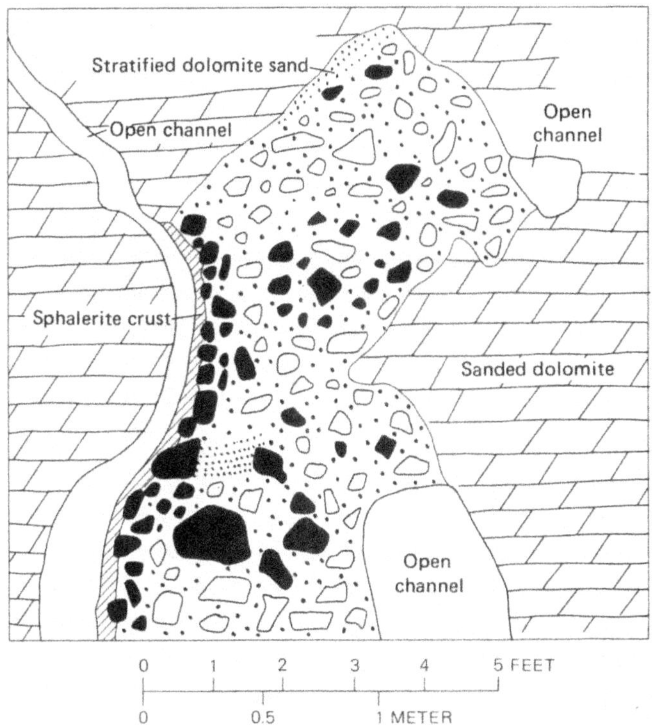

FIGURE 19.—Cave rubble formed during mineralization stage. Rubble contains fragments of siderite and pyrite (black) and is coated by sphalerite along channel at left, 17 level, Eagle mine.

composition as that in unaltered Leadville Limestone. In contrast, the lead in dolomite in contact with ore is of a different isotopic composition and almost identical to the lead in the galena of the ore.

Measurements of radioactivity were made by Hurley (1950, p. 28) on a series of samples of dolomite extending from an ore chimney outward for 800 ft (244 m). No systematic change with distance was evident. The source of the radioactivity was not determined, but it possibly lies in the organic matter in the dolomite. If so the radioactivity values might vary with the kind of dolomite sampled, which is not indicated. Engel and Patterson (1957) reported 0.4 ppm uranium in unaltered Leadville Limestone but did not report the uranium in dolomite.

Wildeman (1970) and Banks (1967) reported a manganese content on the order of hundreds of parts per million in dolomite and tens of parts per million in unaltered Leadville Limestone, indicating some enrichment in manganese during dolomitization. Their determinations, made by X-ray fluorescence methods, are somewhat higher than those indicated in table 3, made by chemical methods.

Roach (1960) reported that porosity of dolomite decreases and thermoluminescence increases with distance from ore. In dolomite of the Leadville, the porosity correlates closely with degree of sanding. In dolomite of the Dyer, porosities of 15–20 percent were measured through a distance of about 10 ft (3 m) adjacent to ore in the absence of visible sanding. From this level, porosity decreases systematically to less than 5 percent at a distance of about 100 ft (30 m) from ore.

Engel, Clayton, and Epstein (1958) made extensive studies of oxygen and carbon isotopic compositions of dolomites in the Leadville to determine whether any systematic change occurs with distance from ore. They found (1) generally lower contents of O^{18} in dolomites than in unaltered Leadville Limestone, (2) generally lower O^{18} contents in the hydrothermal dolomites than in the dolomite of inferred primary origin in the Gilman Sandstone, (3) no O^{18} gradient in the early dark dolomite from the edge of the dolomitized belt to the ore bodies at Gilman, (4) a general but irregular decrease in O^{18} content of other varieties of dolomite in going from the edge of the dolomitized belt toward Gilman, and (5) a decrease in O^{18} through successive crystallization facies. Later, T. S. Lovering and Irving Friedman (unpub. data) analyzed seven samples from different beds of the inferred primary dolomite of the Dyer and found O^{18} contents similar to those of the dolomite in the Gilman Sandstone as reported by Engel, Clayton, and Epstein, ($\delta O^{18} \approx 26$). Seven samples of early dark dolomite in the Leadville, from the same part of the Eagle mine as the Dyer samples, yielded less consistent but generally lower O^{18} values (δO^{18}, 20–24).

Engel and coworkers concluded from the oxygen data that the dolomites of the Leadville recrystallized at higher temperatures than did the unaltered limestone of the Leadville and that the dolomite was of hydrothermal origin. They also concluded that a temperature gradient existed from the outer edge of the dolomitized area (cooler) toward the mineralized center at Gilman (hotter). At the time of their work, δO^{18} values had not been calibrated with a temperature scale, and they supplied no actual temperatures.

T. S. Lovering and Irving Friedman (unpub. data) later calculated approximate temperatures from the oxygen data. They concluded that the early dark dolomite crystallized at temperatures of 150°–175°C and the light-colored dolomite of zebra rock at 200°–300°C. These values are in general accord with the results of limited fluid-inclusion and thermoluminescence studies made by T. G. Lovering (1958, p. 700–704), who reported that the crystallization temperature of dolomite of the pink breccia was 220°–260°C and that the temperatures for zebra rock and pearly dolomite reached a maximum of 330°C.

BLACK CLAY

A greasy black clay occurs abundantly in places in the pink breccia, in the Gilman Sandstone, and at contacts between sulfide ore and dolomite. It also occurs as scattered small clots in sulfide ore and some cave rubble and as microscopic stylolitic veinlets in dolomite rock. The clay is in large part a residual concentrate of the impurities in dolomite rock. Where the clay has formed abundantly, the resulting mixture of clay and residual dolomite closely resembles the "shalified" rock formed by dissolution in mineralized areas of the Upper Mississippi Valley (Heyl and others, 1959, fig. 64).

The clay contains abundant finely divided carbonaceous or bituminous matter and leaves a greasy smudge on the hands. It consists largely of hydromica (illite) and a member of the montmorillonite group whose optical properties suggest beidellite. These materials and accessory quartz, sericite, and dolomite probably were derived from the parent dolomite rock, along with the organic matter. A group of sulfate minerals also present in the clay probably was introduced by the fluids that caused dissolution. This group is represented principally by gypsum, but alunite, magnesium sulfates — epsomite and kieserite — and unidentified iron sulfates have been noted. Finely divided pyrite is present in some of the clay. Wehrenberg and Silverman (1965, p. 333) reported disseminated sphalerite and as much as 15,000 ppm Zn in the clay in at least one locality.

JASPEROID

Dolomites of the Leadville and Dyer have been replaced by jasperoid (cryptocrystalline silica) on a small scale in two areas in the Gilman district, both some distance from the main ore bodies. One area is at the Cave and Coolidge prospect (pl. 1), about 0.6 mi (1 km) northwest of Gilman, where a body of jasperoid about 300 ft (90 m) long is present in the Leadville. The other area is at Red Cliff, where jasperoid occurs in scattered small bodies in the Leadville and Dyer Dolomites and the Gilman Sandstone. Many of these bodies are less than 10 ft (3 m) in greatest exposed dimension. Additionally, Crawford and Gibson (1925, p. 57) noted pieces of jasperoid on the dumps of the Little Chief and Iron Mask mines at Gilman but did not find any exposed in the mine workings.

Much larger bodies of jasperoid occur outside the Gilman district both to the northwest and to the southeast. Because of their remoteness, we doubt that these bodies are directly related to the mineralized center at Gilman. The jasperoid in the northwestern area, shown on the map of the Minturn quadrangle (Tweto and Lovering, 1977, pl. 1) extends from the vicinity of Cross Creek (fig. 1) northwestward about 5 mi (8 km). It is principally in the Dyer Dolomite. The Leadville, which is a limestone in this area, was replaced only locally. The jasperoid of the southeastern area is near Pando (Tweto, 1953; Crawford and Gibson, 1925, pl. 1). In some parts of this area, the entire Leadville, Gilman Sandstone, and Dyer were replaced. The jasperoid replaced zebra rock in both the Leadville and the Dyer, and it replaced collapse breccia that presumably was a product of the sanding or dissolution stage of dolomite alteration. Many varieties of jasperoid are present, and crosscutting relationships among these indicate a complex history of jasperoid deposition.

The jasperoid of all these occurrences has been described in detail by T. G. Lovering (1972, p. 78–81, 99–100), and that of the Cave and Coolidge prospect was also described by T. S. Lovering (1966). T. G. Lovering characterized most of the jasperoid as uncommonly poor in minor elements, though he later found (unpub. data) that certain young jasperoids that vein the older varieties at both Pando and the Cave and Coolidge prospect contain anomalous amounts of silver, tungsten, and some other elements. T. S. Lovering determined from a detailed study of crystal orientations and overgrowths that the solutions that deposited the main body of the jasperoid at the Cave and Coolidge prospect moved generally updip in a convective pattern. He concluded that the source of the solutions was to the northeast and not from the direction of the Gilman ore bodies to the southeast.

We conclude that, at most, deposition of jasperoid was only a subsidiary process in the hydrothermal stage at Gilman. Except for certain of the younger varieties, jasperoid was evidently deposited by waters formed by the mixing and dilution of introduced silica-bearing "hydrothermal" waters with resident ground waters. Deposition of jasperoid may have begun during mineralization, and it probably continued long after. Moreover, the paragenetic evidence would permit two or more distinct periods of jasperoid formation, separated by time gaps of unknown duration.

KARST SILT

The most evident alternation effect in karst silt is bleaching. The silt is normally ocher-yellow to gray, but the silt bodies enclosed in or near sulfide ore are pale green to white. Fragments of dark chert in some parts of the silt are decolorized in rinds a few millimeters thick and evidently were also affected by the bleaching process. The bleached silt contains much more clay than the unaltered silt and, as a consequence, is soft and plastic or pasty. The predominant clays are kaolinite and dickite. These clays are hydrothermal alteration products of an original mixture of illitic clay and other clays which, along with silt-size quartz and chert, compose most of the unaltered silt. Most of the altered karst silt contains pyrite in scattered small grains. We infer that iron oxides that colored much of the original rock were converted to pyrite in the alteration process, though some iron may have been leached from the rock also.

FIGURE 20. — Breccia, clay, and dolomite sand in Gilman Sandstone, 18 level, Eagle mine. *A*, View of area about 6 ft wide and 3 ft high (1.8 × 0.9 m). Light-colored areas at lower left and upper center are brecciated sandstone; dark stratified material at lower right is black clay that contains fragments of sandstone; stratified material in upper left is relithified dolomite sand. *B*, Detail of part of area shown in *A*, showing stratified and crenulated dolomite sand (upper left) cut by a structureless sand "run"; both varieties of dolomite sand are relithified to friable dolomite.

GILMAN SANDSTONE

The carbonate-cemented sandstone and interbedded dolomite of the Gilman Sandstone show pronounced effects of dissolution in the mineralized area at Gilman, and similar effects have been noted at Leadville (Tweto, 1949, p. 182). Products of the dissolution are resedimented quartz and dolomite sands, breccias, and black clay (fig. 20). Some of the resedimented sands differ from those in the Leadville and Dyer Dolomites in being relithified to friable sandstones. This suggests the possibility that the Gilman records two stages of dissolution.

Internal structures in the Gilman are complex. Large slabs of sandstone, some of them as much as 50 ft (15 m) long, have subsided so that they end abruptly against some other material. Some beds of hard, unaltered sandstone grade into structureless friable sandstone that contains fragments of chert and dolomite and pockets and clots of black clay. Dikelets of quartz sand, or of dolomite sand, cut breccias, laminated sands, and relict unaltered sandstone and dolomite. The contact of the Gilman with the underlying Dyer Dolomite is very uneven in places; humps and pinnacles of Dyer rocks extend into the Gilman, and lobes and dikelets of Gilman materials penetrate the Dyer (fig. 21). The contact of the Gilman with the overlying Leadville Dolomite is less irregular, though in places lobes and short dikes of breccia extend from the Gilman into the Leadville. Breccia in the Gilman locally contains fragments derived from overlying strata in the Leadville. A few of the fragments are limestone, though the Leadville is now all dolomite in this area.

The evident intensity of dissolution in the Gilman Sandstone, particularly near ore bodies, indicates that this unit was an important element of the circulation system for the sanding and mineralizing solutions. The sandstone and sedimentary breccia of the Gilman, bounded above and below by unconformities, evidently constituted a particularly permeable zone sandwiched between the dolomite rocks of the Leadville and the Dyer. A possibility exists that some of the permeability was inherited from the period of karst dissolution in pre-Belden (pre-Middle Pennsylvanian) time. At that time, the Gilman was less than 150 ft (46 m) below the surface that was being subjected to karst erosion at the top of the Leadville. It could well have an element of the karst circulation system, but whether it underwent some dissolution is open to interpretation. We attach significance to the fact that some of the resedimented quartz and dolomite sands in the Gilman are relithified, whereas the dolomite sands in the Leadville and Dyer are not. We infer from this, and also from the occurrence of a few limestone fragments — as contrasted to dolomite — in breccias of the Gilman, that some of the dissolution features predate the hydrothermal stage and are of karst origin. If this is so, then why is it that dissolution features in the Gilman seem largely confined to mineralized areas, such as the Gilman and Leadville districts? We postulate that karst dissolution affected the Gilman only locally, in areas of very active karst circulation, and that the channel systems of such areas later provided ready access of mineralizing solutions to the Leadville Dolomite. As a corollary

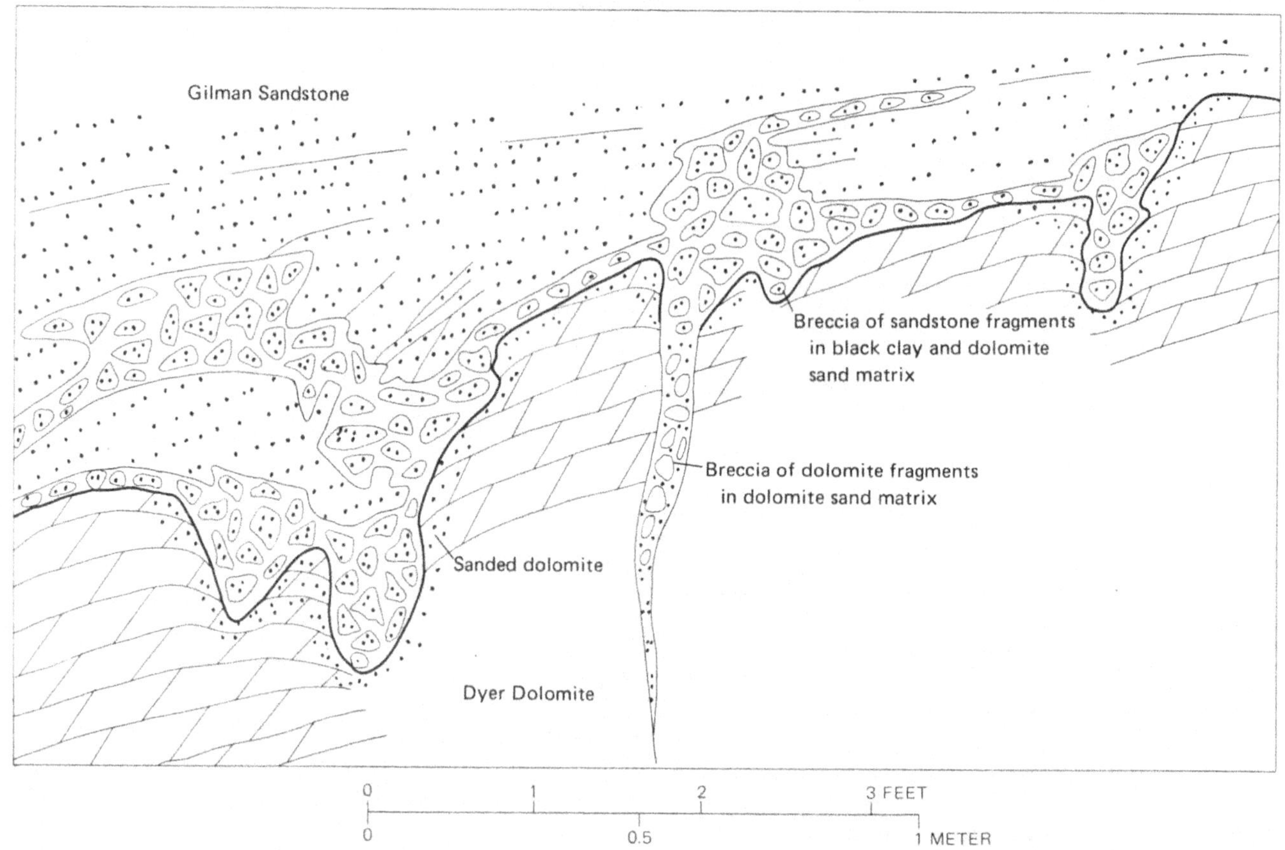

FIGURE 21. — Irregular basal contact (heavy line) of Gilman Sandstone and breccia of the Gilman extending into underlying Dyer Dolomite, 18 level, Eagle mine.

of this postulate, evidence of dissolution in the Gilman might have important application in the search for ore deposits. Much closer studies of the Gilman than have been made previously in the mineral belt are warranted.

SAWATCH QUARTZITE

Like the dolomite rocks of the Leadville and the Dyer, quartzite of the Rocky Point zone in the Sawatch Quartzite was subjected to hydrothermal dissolution on an extensive scale before and, to a lesser extent, during mineralization. Leached and channeled areas are evident throughout the area of mine workings in the quartzite (fig. 11) but are absent in surface exposures to the north and south. Open channels formed by dissolution served both as conduits for mineralizing solutions and as receptacles for some sulfide deposition. Thus, the main areas of mine workings in the quartzite in general coincide with the areas of most intense channeling. The mine workings (fig. 50; pl. 2*A*–*D*) have a seemingly aimless pattern and include many closely spaced long inclines, not because of poor planning, but because many of them follow channelways. Some of the channelways were open and barren and required little work to make into passageways, and others were excavated for the ore in them.

The leaching solutions gained access to the dense quartzite along small faults, joints, bedding planes, and the crackle breccia of the bedding faults in the Rocky Point zone. This is shown by leached zones that spread outward from fractures and bedding planes and by alinement of many of the larger channels in the direction of "slips" or persistent joints. In incipient stages of dissolution, quartzite along fractures is transformed to a spongy form in narrow sheaths enclosing the fractures. Along some fractures, one wall is leached and spongy and the other is entirely fresh and glassy. In the leaching process, the detrital quartz grains were attacked more readily than the siliceous cement; thus, some spongy leached rock consists mainly of the silica cement of the original quartzite.

In more advanced stages of dissolution, open channels formed (fig. 22). The channels range in size from tabular openings less than an inch (2.5 cm) wide along fractures and bedding planes to large anastamosing channel systems hundreds or thousands of feet in length. These large channels are very irregular but are mainly flat openings 1–3 ft (30–90 cm) high and a few feet to tens of feet wide in the plane of the bedding. Some, however, enlarge locally to subcircular openings as much as 20 ft (6 m) in diameter.

A few channels, both large and small, are bordered by concentric curved fractures that indicate spalling. In the Ground Hog mine, some channels enlarge abruptly into large smoothly domed "rooms" walled by the spall fractures. Fractures concentric to these walls divide the quartzite into curved plates that transect the bedding. Small-scale spalling observed in workings off the Forgy incline of the Ground Hog mine (pl. 2*A*) is illustrated in figure 23. One open channel about a foot (≈30 cm) in diameter (fig. 23*A*), in glassy quartzite, is surrounded by a spall zone 2–5 in. (5–13 cm) wide in which the quartzite is split into very thin concentric plates. Fractures between the plates are filled with black sphalerite

FIGURE 22.—Solution channels in Sawatch Quartzite, Rocky Point mine. *A*, Channels along bedding and joints; area shown is about 5 × 8 ft (1.5 × 2.4 m). *B*, Closeup view showing pitting in quartzite along solution channel; area shown is about 1.5 × 1.5 ft (0.45 × 0.45 m).

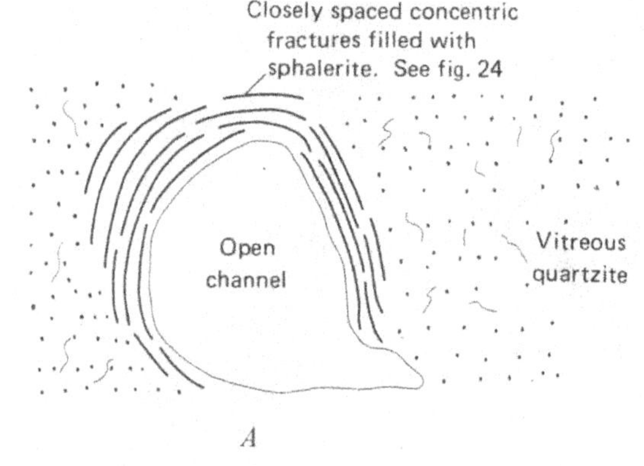

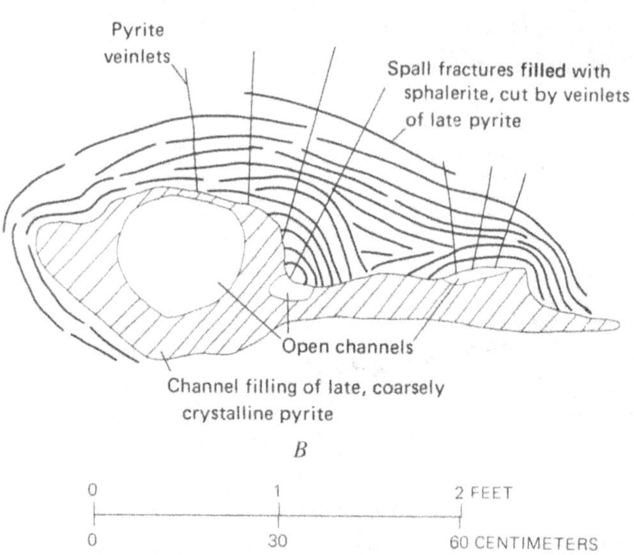

FIGURE 23. — Spall structure around channels in Sawatch Quartzite, Ground Hog mine. *A*, single channel; *B*, merged channels.

(fig. 24). Another channel (fig. 23*B*) has truncated sets of concentric fractures, as if channels had enlarged and merged after initial spalling. This channel is lined with a porous crust of coarsely crystalline pyrite, and fractures concentric to it are filled with black sphalerite.

The fact that spall structure is present in only a small percentage of all the channels observed indicates that the spalling was caused by some specialized mechanism that operated only locally. If the spalling were a static pressure effect, caused by weight of overlying rocks, it should have affected all or most of the channels rather than only a few. Similarly, if it were caused thermally by the introduction of hot fluids into cool rocks, it should also be more widespread. Moreover, the thermal regime in the Gilman district just prior to dissolution and mineralization would scarcely permit cool rocks. (See the section, "Hydrothermal Processes.") Conversely, the sudden introduction of cool fluids into hot rocks is an improbable mechanism because of the widespread evidence of dissolution, which for quartzite demands hot fluids. None of the channels in quartzite, whether with spall structure or not, contain any rubble. Any debris formed by caving or spalling was evidently dissolved and removed in solution, suggesting as the agent a very hot and very active fluid. We conclude that the spalling resulted from local buildups and releases of fluid pressure and accompanying abrupt temperature changes in a circulation system that was undergoing many changes as old conduits were shut off and new ones opened.

Aside from dissolution effects, the quartzite is only moderately altered. The quartzite in the walls of channels and of some fractures generally is slightly argillized or sericitized. Such quartzite is speckled with minute chalky white spots. The width of an altered zone along a channel wall or fracture varies with individual beds of quartzite. The quartzite in some beds is altered to depths of only a few millimeters from the walls of channels or fractures, whereas that in other beds is altered for several meters. Obviously, a wide range in susceptibility to alteration existed among the individual quartzite beds. In the outer parts of altered zones, detrital feldspar grains and some intergranular quartz are replaced by a mixture of halloysite and kaolinite. Closer to channels or fractures, these clays, additional intergranular quartz, and some of the detrital quartz are replaced by hydromica and sericite.

The dissolution of quartzite was accomplished by the same solutions that caused dissolution in dolomite, and it overlapped the early stages of sulfide (pyrite) deposition. (See sections "Summary of Hydrothermal History" and "Hydrothermal Processes.")

FIGURE 24. — Sphalerite veinlets in spall fractures that surround a channel in quartzite. (Compare with fig. 23.)

PRECAMBRIAN ROCKS

Veins in the Precambrian rocks are bordered by altered zones that range in width from as little as 6 in. (15 cm) to as much as 40 ft (12 m). The character of the alteration is generally similar along all the veins but differs in some details along certain veins. The Santa Cruz vein, in the canyon wall about 3,700 ft (1.13 km) south of Gilman (fig. 11), has an altered envelope typical of the fissure veins, though wider than most. This vein is on a strand of the Precambrian Homestake shear zone that was refractured, altered, and mineralized in later — presumably Laramide — time. The country rock beyond an altered zone 40 ft (12 m) wide on each side of the vein is sheared migmatitic granite containing fresh biotite and plagioclase. At the outer fringes of the altered zone, plagioclase is partly altered to allophane, lesser sericite, and minor beidellite. In some of the plagioclase, sericite has developed along the (001) planes, whereas the (010) planes show only allophane. Magnetite in this outermost zone of alteration is largely altered to hematite, and sphene is partly altered to leucoxene. Biotite, potassium feldspar, and quartz are not altered except where microscopic veinlets of sericite or beidellite cross them in zones of microbreccia. At about 25 ft (7.5 m) from the vein, increasing intensity of alteration is marked by the virtual disappearance of plagioclase, which is replaced by a mixture of beidellite and sericite, and by the conversion of biotite to muscovite or coarse sericite. Magnetite, ilmenite, and sphene have disappeared and are represented only by leucoxene and microveinlets of hematite or "limonite." The sericite in this zone is later than beidellite, as shown by replacement relations, and is the dominant alteration mineral present. Closer to the vein the potassium feldspar is sericitized as well as the plagioclase, and the only original minerals left are quartz and apatite. A few feet from the vein, in the zone of most intense alteration, secondary quartz appears; it forms irregular mosaic islands scattered through the groundmass of sericite and is accompanied by some pyrite.

The alteration adjacent to the Ben Butler vein (pl. 2*E*) is similar to but less intense than that along the Santa Cruz vein. The original country rock of the Ben Butler was sheared biotite gneiss characterized by fresh feldspars and partly chloritized biotite. Close to the vein, the feldspars are completely altered, some of them to sericite and others largely to clay. The clay consists of an early halloysite in which kaolinite developed, and the aggregates of these two minerals are replaced to a greater or lesser extent by sericite. Some montmorillonite is present and accompanies the sericite where it replaces the halloysite-kaolinite intergrowth. Near the vein, most of the biotite is altered to clay-sericite aggregates similar to those that replaced the feldspars, though a few flakes of hydrobiotite are present. The sericite that replaced biotite is much coarser grained than that which replaced feldspar, and it is streaked or fringed by thin seams of leucoxene. The abundant magnetite of the original rock is almost completely altered to hematite a few feet away from the vein; next to the vein, both the magnetite and the hematite are partly replaced by pyrite. Some quartz probably was introduced into the rock, but most of the quartz is residual from the original rock.

On the outer fringes of the altered zone along the Whipsaw vein and only a few feet from the vein itself, the biotite of the gneissic country rock is recrystallized to a fine-grained dark-green biotite which has altered along cleavage planes to montmorillonite. Microcline is unattacked; but plagioclase is converted to a fine-grained mixture of kaolinite, fine-grained green biotite, and minor sericite. Magnetite is largely altered to hematite. Closer to the vein, the green biotite is converted to muscovite or coarse sericite, and the kaolinite of the altered plagioclase is changed to beidellite. Adjacent to the vein, leucoxene is converted to fine-grained rutile; pyrite is abundant; and fine-grained quartz is present both in the groundmass of sericite and beidellite and as many small veinlets. The pyrite close to the vein is largely pyritohedral or in cubes modified by pyritohedrons, whereas at a greater distance the pyrite is in cubes and is in smaller grains.

The Star of the West vein in the Tip Top tunnel shows an additional alteration stage marked by late dickite which apparently just preceded the introduction of pyritohedral pyrite. The dickite is in veinlets that cut cleanly across the aggregates of sericite, beidellite, and fine-grained quartz in the altered rocks adjoining the vein.

To summarize, the altered zones adjacent to veins in Precambrian crystalline rocks are noteworthy for the absence of carbonate minerals and chlorite (except chlorite in the sheared rocks of the Precambrian Homestake shear zone). In the outermost zone of alteration, biotite is recrystallized to a fine-grained green biotite, and as the vein is approached, this is altered to sericite or muscovite. Magnetite is changed to hematite in the outer zone, and hematite is replaced by pyrite as the vein is approached. Sphene is replaced by leucoxene, which near the vein is recrystallized to fine-grained rutile. Plagioclase shows the typical sequence of allophane to halloysite to kaolinite to beidellite to sericite with increasing intensity of alteration and proximity to the hydrothermal conduit. Potassium feldspar is relatively resistant but alters to sericite in the zone of most intense alteration. Original quartz is stable throughout the zones of alteration, and near the veins, introduced quartz of two or more generations may be found. Pyrite is cubic near the outer edges of its occurrence and is pyritohedral in the zone of intense alteration close to the vein. Dickite, a higher temperature form of clay than kaolinite, occurs in veinlets cutting sericitized rock and probably just preceded the last-stage, or pyritohedral, pyrite.

The mineralogy of the altered zones indicates that the alteration was accomplished by hot potassium-bearing oxidizing solutions. These were poor enough in calcium, sodium, magnesium, and iron to cause biotite and plagioclase to be unstable. Initially, they had a potassium to hydrogen ratio such that potassium feldspar was stable in them, but with passing time (at any one place), the ratio changed so that the potassium feldspar was replaced by sericite near the veins. Here we see the effects of solutions with a high ratio of potash to hydrogen ion; these solutions were at first in equilibrium with quartz and sericite and later deposited the aluminium silicate dickite and then iron sulfide and the ore minerals. The appearance of dickite at a later stage than the main period of sericitic alteration indicates a marked decrease in the ratio of potassium to hydrogen ion. The presence of dickite suggests a higher temperature than that under which kaolinite formed and a lower pH than that under which sericite had formed earlier.

PANDO PORPHYRY

The sill of quartz latite porphyry, the Pando Porphyry, in the basal part of the Belden Formation is not mineralized and, except in a very few places, is not in contact with ore bodies. However, the sill is not far above the ore bodies in the upper part of the Leadville Dolomite, and it shows alteration effects over and near these ore bodies that seem assuredly to be related to the mineralization. These effects are superimposed on an early, deuteric alteration that affected the porphyry throughout the area. This makes the mineralogic relationships complex.

The deuteric alteration, described in the Minturn quadrangle report (Tweto and Lovering, 1977), produced a sericite-anorthoclase assemblage. Most of the biotite in the rock was destroyed, and plagioclase was partly destroyed. When rock of this character was further altered hydrothermally, the remaining plagioclase was destroyed, calcium and sodium were leached, and much potassium was added to yield a thoroughly sericitized rock. Of the minor components in the deuterically altered rock, allophane and montmorillonite were converted to sericite in the hydrothermal stage, and calcite was destroyed.

Our data, derived from samples in scattered areas, indicate further alteration beyond the sericite stage of hydrothermal alteration in a zonal pattern closely related spatially to the ore bodies. A zonal pattern was also noted by O'Neill, as cited by Radabaugh, Merchant, and Brown (1968, p. 651). Immediately above some ore bodies, the sericitized rock is modified by younger argillic alteration. In this process, kaolinite or dickite and hydromica formed at the expense of sericite and microscopic residual patches of groundmass and phenocrysts of the original (deuterically altered) rock. Additionally, calcite and quartz were introduced as microscopic veinlets and replacements in porphyry over some of the manto ore bodies. Pyrite and sparse galena and sphalerite accompany the calcite and quartz, indicating that this stage of alteration occurred during sulfide deposition. No calcite was detected in the argillized sericitic rock over chimney ore bodies, but kaolinite, pyrite, and quartz were found to be abundant.

In summary, potassic hydrothermal alteration in the general area of the ore bodies greatly increased the degree of sericitization in porphyry that had already been partly sericitized deuterically. Within the sericitized body and in close spatial relation to ore bodies are areas in which a late argillic alteration is superposed on the sericitic alteration. The argillic alteration indicates a change to potassium-poor acid solutions. These solutions also added quartz, calcite, and pyrite to the argillized rock and are correlated with ore solutions.

MINERALS AND PARAGENESIS OF THE ORE DEPOSITS

The minerals pyrite, siderite, and sphalerite constitute a very large part — perhaps 90 percent — of the total volume of sulfide ore deposits in the Gilman district. A measure of this preeminence is the fact that zinc from the sphalerite has accounted for almost two-thirds of the total value of production from the entire district (table 2). The three minerals occur mixed and also in essentially monomineralic bodies. In mixed form, they constitute the manto ore bodies, in which they are accompanied by some galena and traces of chalcopyrite. Pyrite occurs additionally as the cores of the chimney ore bodies, and, as branches of these bodies, it forms cores of the lower ends of some of the manto ore bodies (fig. 25). In parts of the chimney bodies, the pyrite is pervaded by volumetrically minor but economically very significant copper, silver, gold, and lead minerals that transform it to copper-silver ore. The central pyrite bodies of the chimneys are surrounded by discontinuous shells of zinc ore, parts of which consist almost entirely of sphalerite. In turn, the

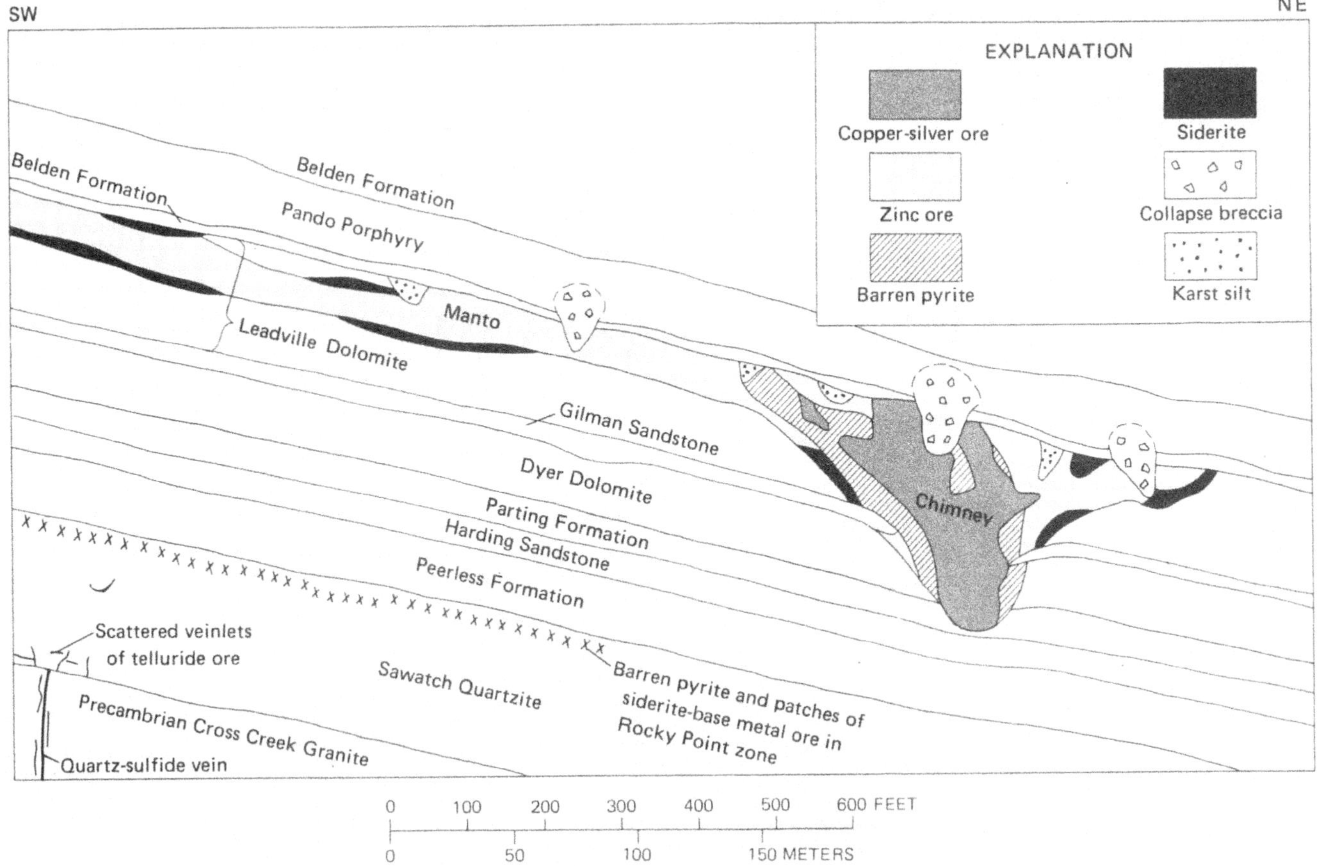

FIGURE 25. — Distribution of materials in manto and chimney ore bodies (patterned) and stratigraphic setting of ore bodies (diagrammatic). Modified after Radabaugh, Merchant, and Brown (1968, fig. 5). Vertical and horizontal scales are equal.

zinc or sphalerite shell is surrounded in the upper parts of the chimneys by a discontinuous shell of siderite. Similar shells of siderite also surround the manto ore bodies in many places.

The mineralogy and depositional history of both pyrite and siderite are complex because slightly different varieties of these minerals were deposited at different times. Pyrite paragenesis is further complicated by evidence that some iron sulfide first existed as marcasite and later was transformed to pyrite.

In discussion that follows, we shall first consider pyrite, siderite, and marcasite because these minerals provide a basic framework to which the depositional history of all other minerals can be related. We divide the discussion under three main headings. One discussion deals with the base-metal mineral assemblages of the manto ore bodies, the chimney ore bodies prior to enrichment by the minerals of the copper-silver ores, and with most of the sulfide deposits in the Sawatch Quartzite and the Precambrian rocks. A second discussion deals with the paragenetically — and perhaps geologically — younger group of minerals in the copper-silver ores of the enriched parts of the pyrite chimneys and with what we judge to be correlative components in the deposits in the Sawatch Quartzite, the Precambrian rocks, and in the Leadville Dolomite at Red Cliff. A third grouping deals with the minerals of the oxidized ores. Under each of the headings, the ore minerals are discussed first and then the gangue minerals other than the early pyrite and siderite.

EARLY BASE-METAL SULFIDE ORES

PYRITE

Pyrite (FeS_2) is the most abundant metallic mineral in the ore deposits of the Gilman district (fig. 26). It is the chief component of the chimney ore bodies, of the ore deposits in the Sawatch Quartzite where unoxidized, and of the veins in Precambrian rocks; it is also a major component in the manto ore bodies. The pyrite has many forms: coarsely crystalline, massive, sandy or granular, porous and spongy, banded, bladed, or finely disseminated in associated sulfide and gangue minerals, quartzite, or Precambrian rocks. The bladed variety (fig. 27) is a peculiar form for pyrite and is thought to be pseudomorphous after marcasite.

We distinguish pyrites of three distinct paragenetic ages, or age groups, among the ore deposits and shall

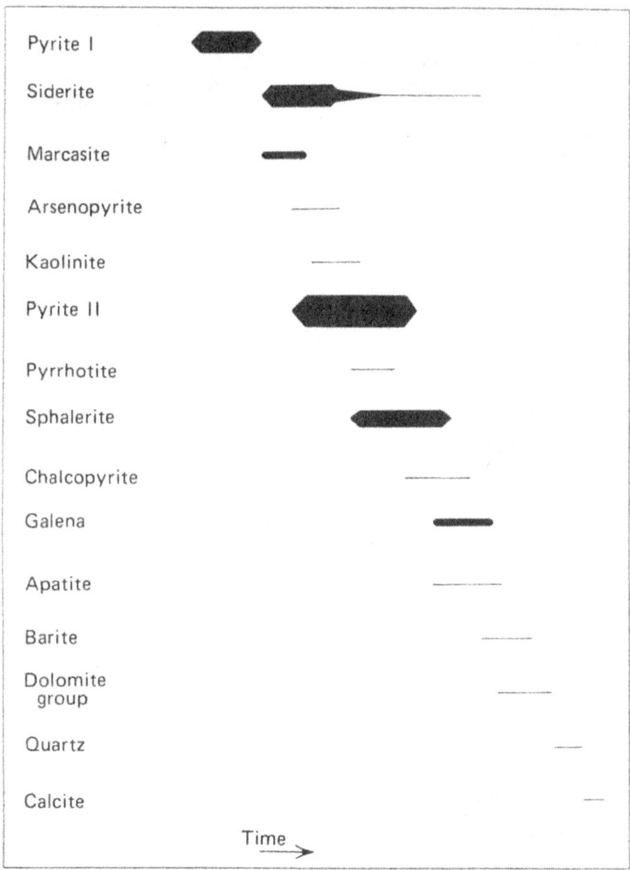

FIGURE 26. — Paragenetic sequence and relative abundances of minerals in early base-metal sulfide ores. Widths of bars are proportional to abundance.

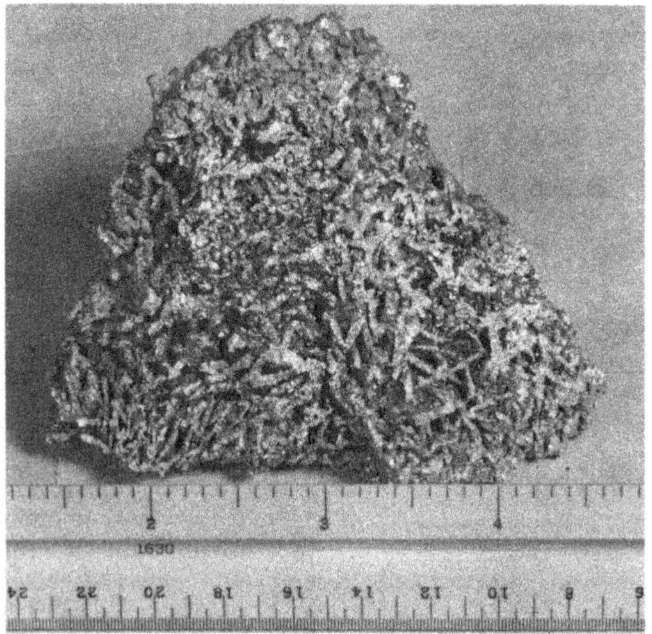

FIGURE 27. — Bladed texture in pyrite. This texture suggests that the pyrite is pseudomorphous after marcasite. Compare with figure 30.

refer to these as pyrites I, II, and III in contexts where distinction is pertinent. Our data on the distribution of these three pyrites are skimpy because at the time of our field observations and sampling we were unaware of the complexity of the pyrite paragenesis.

As judged from our limited data, pyrite I (oldest) occurs in the ore bodies in dolomite rock only in the cores of the chimneys and in the pyrite cores in the lower parts of the mantos. In the Sawatch Quartzite, it is the widespread disseminated pyrite and the pyrite of most fracture fillings. In the Precambrian rocks, it probably is the disseminated pyrite associated with quartz in the inner parts of the altered zones, and it may be the pyrite of some veins that contain no base metals. Pyrite I was deposited before siderite. This is best shown in the Sawatch Quartzite, where ore consisting of siderite and associated pyrite II (or marcasite), sphalerite, galena, and chalcopyrite coats and veins the pyritized quartzite of the pyrite I stage.

Pyrite II is closely associated with siderite and sphalerite in the paragenetic sequence siderite–pyrite II–sphalerite–chalcopyrite–galena (fig. 26). This sequence is characterized by overlapping age relationships, and small amounts of pyrite II are as young as galena. Pyrite II is intergrown with pyrite I in parts at least of the pyritic cores of the chimney ore bodies, but its chief occurrence in the chimneys is in association with the sphalerite and siderite of the outer shells. Pyrite II is the chief — if not the only — pyrite of the sphalerite-bearing parts of the manto ore bodies. In these ore bodies, it is intimately intergrown with sphalerite (fig. 28). Some of it is in the bladed form, which is interpreted to indicate former marcasite (fig. 27). In the ore bodies in Sawatch Quartzite, pyrite II is associated with siderite and accompanying sulfide minerals. In the veins in Precambrian rocks, siderite is absent and pyrite II is inferred to be the pyrite associated with sphalerite, chalcopyrite, and galena.

Pyrite III is a late pyrite associated with the copper-silver ores in the chimney ore bodies. A paragenetically late, coarsely crystalline auriferous pyrite, which is present locally in the ore deposits in the Sawatch Quartzite and the Precambrian rocks, is also tentatively assigned to pyrite III. Pyrite III is discussed with the minerals in the copper-silver ores.

Differences exist between pyrites I and II and, depending upon locality, even within these groups. As seen in polished sections, pyrite I from the chimney ore bodies is slightly darker and harder than pyrite II. It is characterized by abundant tiny inclusions of dolomite, which are interpreted to be relict from the dolomite rock replaced by the pyrite; it also contains inclusions of quartz. The grains are small and equant and show traces of cubic or pyritohedral faces. Nitric acid stains

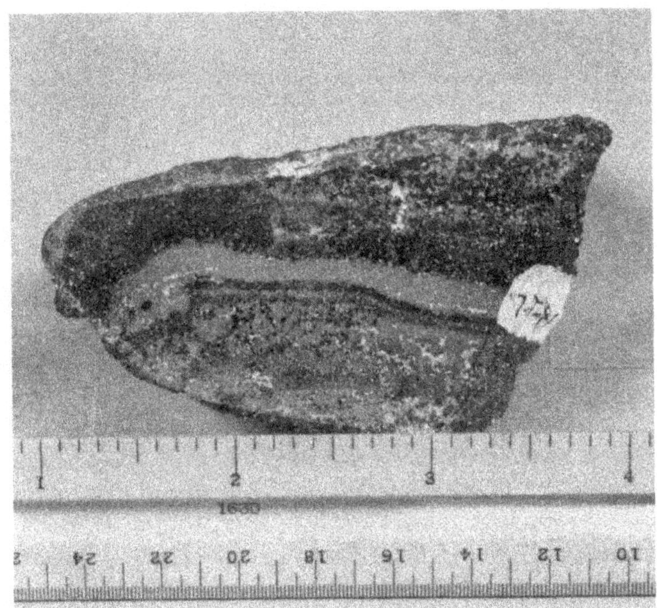

FIGURE 28. — Banded sphalerite (dark) and pyrite (light), typical of much of the zinc ore in the mantos.

the pyrite I light to medium brown in irregular areas. Inclusions in pyrite II are principally siderite and ore sulfides rather than dolomite. The grains are irregular and some are overgrowths on pyrite I. Pyrite II is etched but not stained by nitric acid.

In the deposits in Sawatch Quartzite, differences exist in the crystal habit of pyrite I in belts that extend approximately down the dip of the strata. In one belt the pyrite may be pyritohedral; in an adjoining belt it may be cubic or dodecahedral, or it may show distinctive modification by octahedral, diploid, or trisoctahedral faces. Pyrite crystals in the Percy Chester mine are grating to the touch, owing to the presence of minute wire-edge ridges that follow the complex striations and lines of zonal growth on the crystal surfaces. Crystals from most other localities are smooth to the touch. The variations in crystal habit almost certainly reflect differences in chemistry of ore solutions or conditions of deposition. They justify further study, but we did no more than observe them and did not attempt to map them.

Pyrite in different localities or occurrences differs markedly in susceptibility to alteration. This difference seems to correlate with the nature and abundance of inclusions in the pyrite, though differences in the compositions of various pyrites probably are also a factor. Pyrite I in the Sawatch Quartzite contains only quartz inclusions, and it tends to remain bright and solid after long exposure in mine workings or on dumps. Pyrite I in the dolomite rocks, which contains dolomite inclusions, tarnishes but does not crumble. In contrast, most pyrite II tarnishes and crumbles rapidly upon exposure, presumably from the presence of inclusions of other sulfide minerals, such as sphalerite, chalcopyrite, and galena.

Quantitative spectrographic analyses were made to determine whether chemical differences exist between pyrites I and II or between resistant and nonresistant varieties. The results were inconsistent, probably owing to failure to eliminate all inclusions from the analyzed materials. In general, however, pyrites from deposits in the Sawatch Quartzite, including examples of pyrites I, II, and possibly III, were found to be poorer in minor elements than those in dolomite rocks. Pyrites from the quartzite setting showed a maximum of 0.25 percent total copper, zinc, and lead, whereas those from the dolomite setting showed higher and widely varying contents of these metals and no evident correlations with any type of pyrite. All samples analyzed contained arsenic in the range of 0.05–0.3 percent and selenium in the range of 0.0001–0.0024 percent. Iron-sulfur ratios are very near the theoretical ratio of 0.872 in about half the samples of all three types of pyrite; the remainder has higher ratios, ranging from 0.872 to 0.897.

SIDERITE

Siderite ($(Fe,Mn,Mg)CO_3$) is the chief gangue mineral of the sulfide ores in dolomite rocks and of the base-metal-bearing ores in the Sawatch Quartzite. It has not been observed in the veins in Precambrian rocks. In the oxidized parts of ore bodies in both dolomite and quartzite, its former presence is readily recognized from abundant pseudomorphs of its distinctive crystals in aggregates of iron and manganese oxides.

Most of the siderite is manganiferous (table 4) and hence is often referred to as manganosiderite. It is also magnesian. Radabaugh, Merchant, and Brown (1968, p. 655) reported that the siderite averages 28 percent Fe, 10 percent Mn, and 8 percent MgO and that ranges in individual samples are 18.9–35.7 percent Fe, 6.5–16.5 percent Mn, and 5.3–12.1 percent MgO. They noted that the manganese content is lowest where the sulfide-mineral content of ore is highest. These data evidently refer to the siderite in deposits in dolomite rocks. At least some siderite in deposits of the Sawatch Quartzite is nearly devoid of manganese despite a pinkish-gray color that suggests a manganese content (table 4, No. 66T46).

Some of the siderite in the outer shells of the chimney and manto ore bodies is a dense light-gray finely crystalline variety in which structures of the original dolomite, such as bedding and zebra-rock banding, are preserved. This siderite is devoid of sulfide

TABLE 4. — *Manganese and minor element content of siderite, Eagle and Champion mines, Gilman district*

[Quantitative spectrographic analyses by J. D. Fletcher, U.S. Geological Survey. Values are in percent. Elements also present in all samples as traces but which are below sensitivity limits are Bi, <0.004; Cd, <0.02; Ga, <0.001; Ge, <0.002; In, <0.01]

Sample No. Field	Sample No. Laboratory	Location	Mn	Ag	Ni	Pb	Sn
9T41	D1455	No. 1 manto, 14 level.	>10.0	0.0016	<0.0004	0.012	0.004
60T46	D1459	No. 1 manto, 15 level.	9.1	.0012	<.0004	.024	.002
8T41	D1454	No. 2 manto, 14 level.	7.7	.00034	.0006	.035	.003
28T41	D1457	No. 2 manto, 15 level.	8.3	.0036	<.0004	.18	.005
66T46	D1458	Champion mine, in quartzite.	.048	.00046	.0010	.012	<.002

minerals, and it closely resembles some varieties of the dolomite wallrocks. Its composition evidently varies with locality, for the refractive index n_O is different in each of four samples and ranges from 1.802 to 1.823.

Siderite that is intergrown with pyrite II and sphalerite is generally coarsely crystalline and porous. This siderite seems to be in part a recrystallization product of the fine-grained variety and in part a coprecipitate with pyrite II and sphalerite. The refractive index n_O ranges widely, from 1.805 to 1.840 as determined from a large number of samples. No correlation between index value and location in ore bodies was noted, except that the siderite intergrown with the bladed pyrite that is inferred to be pseudomorphous after marcasite is generally of high refractive index. This high index suggests a high iron content.

Late siderite occurs in abundance as coarse crystals in vugs in the manto ore bodies (fig. 29). These crystals have a characteristic lens or cockscomb shape and are light yellow gray to pale brown. They are in general of lower refractive index than the siderite of other occurrences: n_O, 1.807–1.823. The late siderite shows various relations to chalcopyrite and galena and evidently is not all of the same paragenetic age. In some vugs, siderite rests on chalcopyrite and is overlain by galena. In others, siderite crystals rest on galena.

FIGURE 29. — Rosettes of late siderite on black sphalerite, showing characteristic lens-shaped crystals. Sphalerite ore shows typical coarse crystalline and finely vuggy character.

Siderite in the ore deposits in Sawatch Quartzite occurs mainly as a coating on the walls of cavities and as veinlets in fractures; it replaces quartzite on a minor scale. The siderite of replacement origin is in felted aggregates of tiny tabular crystals surrounding microscopic remnants of quartzite. The siderite in the quartzite setting is clearly younger than pyrite I and is contemporaneous with marcasite, as discussed under that heading. It is paragenetically older than pyrite II and the sphalerite, chalcopyrite, and galena that accompany it in minor amounts.

MARCASITE

Marcasite (FeS_2) is preserved abundantly in only one locality in the district — in the southeastern workings of the Percy Chester mine in the Sawatch Quartzite (pl. 2*C*). However, at least part of the pyrite II in the manto ore bodies in dolomite rocks has the texture of coarsely crystalline marcasite and evidently existed previously as marcasite. Pyrite II of this character is closely associated spatially with other varieties of pyrite II that show little or no suggestion of a preexistence as marcasite. Conditions of temperature and pH during deposition of the iron sulfides evidently were very near the borderline between the pyrite and marcasite stability fields.

The marcasite in the Percy Chester mine occurs as a coating or filling in solution channels in quartzite that contains small amounts of disseminated pyrite I. Some of the marcasite is in mats of coarse cockscomb crystals (fig. 30). Siderite in minor amounts is intergrown with the marcasite; scattered crystals of arsenopyrite (fig. 31), sphalerite, and chalcopyrite, and patches of kaolinite rest on the marcasite. Marcasite and siderite are reciprocally distributed in the channel system. Marcasite decreases in abundance updip as siderite increases. As judged from crystal pseudomorphs in iron oxides, marcasite originally extended updip only a short distance beyond the base of the present oxidized zone. Conversely, in the deepest workings marcasite is present to the exclusion of siderite.

Marcasite is also present in intergrowths with siderite in a few small areas in the Rocky Point and Ground Hog mines and was observed on the dump of the Bleak House mine. These mines are all in the Sawatch Quartzite.

Marcasite was observed microscopically in two ore samples from the deposits in dolomite rock, one from near the bottom of the Lower No. 2 chimney and one from the outer part of the zinc shell of the 4-5 chimney on 17 level (fig. 11). In both occurrences, tiny needles and blades of marcasite encrust pyrite grains, but we are uncertain whether these crystals are pyrite I or pyrite II. If they are pyrite II, the marcasite signifies a local repetition in part of the paragenetic sequence. The sample from the 4-5 chimney also contains an isotropic variety of marcasite. This variety forms irregular bodies that enclose grains of pyrite and blades of normal marcasite, and it is cut by veinlets of siderite. It has a pale-brown tint in polished section and is softer than normal marcasite. According to R. P. Marquiss of the U.S. Geological Survey (oral commun.) the powder X-ray pattern matches that of marcasite except for a faint line at "d" 2.20. Spectrographic analysis showed 0.13 percent copper in addition to iron and sulfur.

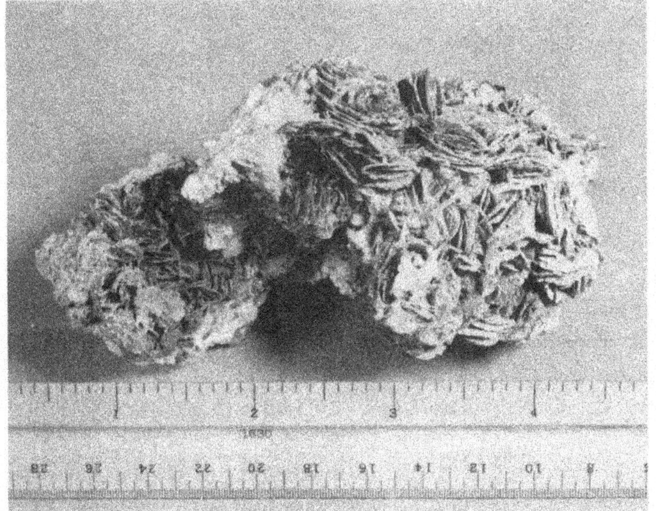

FIGURE 30.—Marcasite crystals from Percy Chester mine. White coating in local patches is kaolinite.

SPHALERITE

Sphalerite ((Zn,Fe)S) is the principal ore mineral of the Gilman district. Almost all that has been mined has come from the manto ore bodies and the outer shells of the chimney ore bodies in dolomitic rocks. The mineral is also a constituent of the sideritic ores in quartzite and of the fissure veins in Precambrian rocks, but, except in the Bleak House mine, it is not abundant enough in these occurrences to be of economic significance.

The sphalerite of the mantos, the chimney shells, the sideritic deposits in quartzite, and of parts of the veins in Precambrian rocks is the black iron-bearing variety marmatite. A different, paragenetically younger, light-colored and iron-free variety of sphalerite is a minor accessory mineral in the copper-silver ores of the chimneys and occurs also in small amounts in some of the veins in Precambrian rocks. This variety has contributed only negligibly — if at all — to the zinc output of the district.

The characteristic black sphalerite of the Gilman district is varied in composition. As reported by Radabaugh, Merchant, and Brown (1968, p. 655), it averages about 54 percent zinc, 11 percent iron, 0.3 percent copper, and 0.2 percent each of manganese, cadmium, and lead. The iron content ranges from 6.5 to 11.8 percent and decreases generally from the lower ends of the mantos toward the upper ends. However, the iron content differs among samples from the same area and also differs by as much as 6 percent between growth zones in a single crystal (Radabaugh and others, 1968, p. 655). As seen in polished section, much of the sphalerite contains blebs of exsolved chalcopyrite, which probably is the main occurrence of the copper reported in both the sphalerite and the manto ores. The amount of exsolved chalcopyrite present in the sphalerite was observed to vary widely in different samples, but we have no data on a pattern of variation.

The black sphalerite also contains significant amounts of cadmium and silver, as well as a little lead (table 5). The silver contents of the samples listed in table 5 translate to a minimum of 3 ounces silver per ton of pure sphalerite. This content would account for half or more of the 1.0 ounce of silver per ton reported in average zinc ore at Gilman (Radabaugh and others, 1968, p. 659). What part, if any, of the silver in the sphalerite is eventually recovered after smelting of zinc concentrates is not known to us.

Black sphalerite is paragenetically younger than pyrite II, but it shows an overlapping relationship with this pyrite. As seen in polished sections, the pyrite II and sphalerite of some intergrowths show mutual boundaries that suggest simultaneous deposition. In other,

TABLE 5. — *Minor element content of sphalerite, Eagle and Ground Hog mines, Gilman district*

[Quantitative spectrographic analyses by J. D. Fletcher, U.S. Geological Survey. Values are in percent]

Sample No. Field	Sample No. Laboratory	Location	Ag	Bi	Cd	Ga	In	Mn	Ni	Pb	Sn
		Black sphalerite:									
60T46	D1443	No. 1 manto, 15 level	0.014	0.006	0.26	0.012	0.028	0.20	<0.0004	0.24	0.007
32T41	D1437	4-5 chimney (shell), 17 level.	.0042	<.004	.06	.001	.011	.14	.0066	.064	.002
70T40	D1434	Lower No. 3 chimney, 20 level.	.068	.005	.26	.008	.017	.33	<.0004	.19	.018
49T46	D1442	Sideritic ore in quartzite, Ground Hog mine.	.016	<.004	.48	.014	<.002	.084	<.0004	.22	.044
191T41	D1441	Light sphalerite, 18-35 chimney, 20 level.	.030	.012	.38	.016	.56	.080	<.0004	.13	.082

more typical intergrowths, the sphalerite embays, veins, and coats pyrite II, as well as the still older siderite and pyrite I.

CHALCOPYRITE

Chalcopyrite ($Cu_2Fe_2S_4$) is a minor component in the sphaleritic manto ores in dolomite rocks and a common component in the sideritic ores in quartzite and in the fissure veins in Precambrian rocks. In the manto ores, it is observed macroscopically only as occasional small crystals resting on sphalerite and pyrite. It is seen microscopially as exsolution blebs in sphalerite, as tiny veinlets cutting some grains of sphalerite, and as sparse interstitial particles in pyrite-sphalerite intergrowths.

In the sideritic deposits in quartzite, chalcopyrite is fairly common, but, except in local pockets, it is not abundant enough to make minable the unaltered sideritic material. In the zone of secondary sulfide enrichment, however, copper leached from chalcopyrite in the oxidized zone is added as chalcocite to that in chalcopyrite, making small ore bodies that are valuable principally for copper and silver.

Chalcopyrite is relatively abundant in the fissure veins in Precambrian rocks, and its contained copper added materially to the value of the gold-silver-copper-lead ores mined from small ore shoots in these veins. Sulfide ore of the Mabel mine, the most productive of the veins in Precambrian rocks, contained as much as 3 percent copper, presumably as chalcopyrite (Crawford and Gibson, 1925, p. 67).

The chalcopyrite of the copper-silver chimney ores is discussed separately under that heading.

GALENA

Galena (PbS) is an important accessory component of the sphaleritic manto ores in dolomite rocks and of the veins in Precambrian rocks. It is only a minor component of most of the sideritic deposits in quartzite but is an important one in the Bleak House ore body in quartzite. The galena in the mantos occurs principally in pockets scattered unevenly through the zinc ore. In the No. 1 manto, a string of pockets or a long ribbon of galena ore twists around the border zone of the zinc ore body through most of its length. In all the mantos, the proportion of galena to sphalerite increases irregularly updip. As judged by the abundance of the lead minerals cerussite and anglesite in the oxidized parts of the mantos, galena must have been abundant in the uppermost parts of the mantos.

The galena in the mantos occurs principally as cubic crystals in vugs in sphalerite-pyrite-siderite ore. In large vugs, the galena crystals exceed an inch ($\approx$3 cm) in size, though in most places they are smaller. Galena occurs also as microscopic veinlets in sphalerite and as minute grains in pyrite.

The galena of the manto ore bodies, and also the little that is present in the sideritic deposits in quartzite, contains a significant amount of silver, though far less than the galena of the copper-silver ores in the chimney ore bodies (table 6). If the sample reported in table 6 is typical, the manto galena contains about 15 ounces silver per ton. As the average manto ore contains 2.0 percent lead, or 2.3 percent of galena, the silver in galena amounts to 0.35 ounces per ton of average zinc ore; this silver is presumably recovered upon smelting of the lead concentrate.

PYRRHOTITE

Pyrrhotite ($Fe_{1-x}S$) is present in very small amounts in parts of the mantos and chimneys in dolomite rocks and in the veins in Precambrian rocks. It has not been observed in the deposits in quartzite. In the ore bodies in dolomite rocks, it is found principally in the highly pyritic lower ends of the mantos and in the contiguous pyritic cores of the chimneys. In these occurrences, it is seen microscopically as small anhedral grains interstitial to pyrites I and II and black sphalerite. Some of the grains show replacement relations against siderite and are in turn embayed by chalcopyrite. Pyrrhotite observed in samples from the veins in Precambrian rocks is younger than a pyrite thought to be pyrite II and older than black sphalerite and chalcopyrite.

TABLE 6. — *Minor element content of galena, Eagle and Ground Hog mines, Gilman district*

[Quantitative spectrographic analyses by J. D. Fletcher, U.S. Geological Survey, except as noted. Values are in percent]

Sample No. Field	Sample No. Laboratory	Location	Ag	Bi	Cd	Ga	Ge	In	Mn	Sb[1]
43T41 . . .	D1428 . . .	No. 1 manto, 15 level.	0.053	0.020	<0.02	<0.001	<0.002	<0.002	0.0015	0.0075
174T41 .	D1432 . . .	18-35 chimney, 18 level.	[2]4.0	.55	<.02	<.001	<.002	<.002	.0013	.025
44T46 . . .	D1429 . . .	Sideritic ore in quartzite, Ground Hog mine.	.094	.048	<.02	<.001	<.002	<.002	.004	.03

[1] Colorimetric analyses by C. E. Thompson.
[2] Sample contained inclusions of hessite, argentite, and freieslebenite. Analysis by J. J. Fahey.

ARSENOPYRITE

Arsenopyrite (FeAsS) occurs very sparsely in the manto ore bodies in dolomite rocks and in the sideritic ore deposits in quartzite. It is seen in some polished sections of manto ores as small anhedral grains that seem to be contemporaneous with pyrite II and older than black sphalerite. As seen in a sample of sideritic ore from the Percy Chester mine in quartzite, arsenopyrite occurs as tiny crystals that coat marcasite (fig. 31).

MAGNETITE AND HEMATITE

The iron oxides magnetite (Fe_3O_4) and hematite (Fe_2O_3) were not observed by us in the sulfide ores, but they are reported to be present locally in ore deposits of unspecified character in quartzite, presumably in the Rocky Point mine (Radabaugh and others, 1968, p. 654).

KAOLINITE AND DICKITE

Small pockets of white clay are scattered sparsely through all the base-metal sulfide deposits. They are generally associated with pyrite II or marcasite (fig. 30), and some have pyrite crystals scattered through them. As tentatively identified from optical properties and differential thermal analyses, the clay is principally kaolinite or dickite ($Al_2O_3 \cdot 2SiO_2 \cdot 2H_2O$) but contains some residual halloysite. The clay is evidently of hydrothermal origin and near pyrite II and marcasite in the paragenetic sequence. Some of the clay in the ore deposits in quartzite is contaminated with illite and sericite that probably were derived from the quartzite.

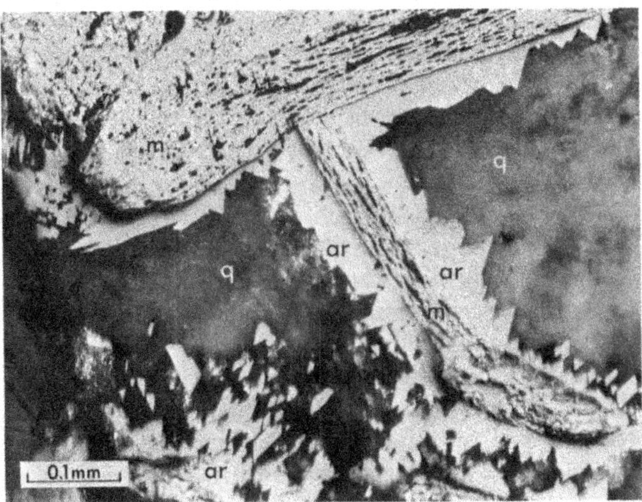

FIGURE 31. — Arsenopyrite (ar) coating marcasite (m), in ore in quartzite, lower workings of Percy Chester mine; q, quartz.

APATITE

Small prismatic crystals of nearly colorless apatite ($Ca_5(PO_4)_3(OH)$) are present in scattered vugs in the manto ore bodies in dolomite rocks, and some have been observed microscopially in intergrowths with black sphalerite. A younger and distinctively different green apatite occurs in the copper-silver ores of the chimney ore bodies in dolomite rocks. This apatite, which is in the form of short striated prismatic crystals with prominent basal pinacoids, occurs in vugs, generally with chalcopyrite. The refractive indices (n_O) of both apatites are in the range 1.646 to 1.653, which suggests hydroxylapatite. The green apatite is anomalously biaxial with $2V$ near 20°.

BARITE

Barite ($BaSO_4$) is a paragenetically late mineral that is widely but sparsely distributed through the ore deposits in dolomite rocks and in quartzite. In both settings it occurs as clear, honey- to amber-colored tabular crystals, 5–10 mm in greatest dimension. Many of the crystals show growth zones that are distinguishable by slight differences in color. These zones are asymmetric and are widest on the sides that face the downdip direction of the ore bodies, suggesting that the flow of solutions came from that direction. The barite crystals are found only on vug surfaces, and in different places they rest on siderite (including late vug-siderite), pyrite, sphalerite, and galena. In the quartzite setting, barite occurs only in the sideritic deposits. It survived oxidation of these deposits and thus proved useful in tracing the courses of original siderite channelways through the masses of iron and manganese oxides in the oxidized zone (pl. 2*D*). In the oxidized zone, barite that rests on psilomelane was observed to be colorless and, in general, to lack growth zones. This variety may have recrystallized during oxidation of the ore deposits.

DOLOMITE GROUP

Species of the dolomite group ($Ca(Mg,Fe,Mn)(CO_3)_2$) occur as paragenetically late crystals or drusy coatings in vugs in the ore bodies in dolomite rocks and very rarely in the sideritic deposits in quartzite and in the veins in Precambrian rocks. The crystals are typically saddle shaped, and some are as large as 5 cm across. They are among the youngest species in the base-metal ore deposits (fig. 26), and in places they even coat barite. The crystals differ in composition from place to place. Most of them contain 3–4 percent each of iron and manganese (Radabaugh and others, 1968, p. 654) and are intermediate between dolomite, ankerite, and kutnahorite. Some, however, are very near dolomite in composition, as shown by refractive indices (n_O) in the range 1.68–1.69.

QUARTZ

Quartz (SiO_2) occurs only in very minor quantity in the ore deposits in dolomite rocks. It is typically seen as slender crystals a few millimeters long projecting into interstitial openings in porous sulfide ore or siderite. Where present in vugs that contain late siderite and species of the dolomite group, it rests on these minerals. Microscopic veinlets of quartz were observed to cut the sulfide minerals in a few polished sections of ore samples.

The chief occurrences of quartz in the base-metal deposits is in the veins in Precambrian rocks, where it is the most abundant of the constituents. The quartz in the vein fillings is coarse grained. It contrasts in this respect with the fine-grained quartz in the silicified wallrocks, formed in place by hydrothermal alteration processes. Deposition of vein quartz evidently both preceded and accompanied sulfide deposition. The quartz is cut by veinlets of sulfide minerals, but in turn, veinlets of quartz cut the sulfide aggregates in places.

CALCITE

Calcite ($CaCO_3$) is a rare mineral in the sulfide deposits at Gilman. It has been observed only in the ore bodies in dolomite rocks, where it occurs in a few vugs and rarely as microscopic veinlets cutting sulfide minerals. The crystals in vugs are scalenohedral in habit and generally are poorly formed. They rest on crystals of the dolomite group and probably are the youngest of all the hypogene minerals in the deposits.

COPPER-SILVER AND RELATED ORES

Copper-silver ore occurs principally in distinct but very irregular bodies within the pyrite cores of the chimney ore bodies in dolomite rocks (fig. 25). Small bodies are also present in places within the manto ore bodies (fig. 11). Copper-silver ore is characterized by the elements copper, silver, gold, lead, antimony, bismuth, and tellurium. The principal ore minerals are argentiferous and auriferous chalcopyrite, tetrahedrite, and galena. Some silver and gold occur as tellurides, principally as inclusions in galena. On the basis of their richness in precious metals, we relate certain deposits other than those in the chimneys to the copper-silver stage of mineralization: (1) the isolated veinlets of telluride minerals in the Sawatch Quartzite; (2) late, coarsely crystalline auriferous pyrite present locally in the ore deposits in quartzite and in the veins in Precambrian rocks; and (3) the silver-rich galena-sphalerite deposits in the Red Cliff area.

The minerals of the copper-silver ores differ in habit of occurrence as well as in their compositions from those of the sphalerite-pyrite-siderite ores. They occur principally as encrustations, vug fillings, and veinlets in the host pyrite of the sulfide chimneys. Some — notably chalcopyrite — also replaced pyrite extensively. The copper-silver ores are clearly products of a later mineralization stage than the sphalerite-pyrite-siderite ores. The question of how much later is discussed in the section on age of mineralization.

The paragenetic sequence and relative abundances of ore minerals in the copper-silver ores, exclusive of the host pyrite, are illustrated in figure 32.

CHALCOPYRITE

Chalcopyrite ($CuFeS_2$) is the chief copper mineral in the copper-silver ores of the chimney deposits in dolomite rocks. The chalcopyrite of this occurrence extensively replaced the older, massive pyrite (pyrites I and II) of the chimney bodies and also was deposited as a filling in voids in porous pyrite and as an encrustation in vugs. Where black sphalerite and siderite are present in the pyrite bodies, as at their borders, chalcopyrite shows similar replacement and filling relations to them. Although some of the chalcopyrite in the chimneys could have been deposited at the pyrite-siderite-sphalerite stage of mineralization, most of it is distinctly younger.

Chalcopyrite is closely associated with tetrahedrite and the little bornite (Cu_5FeS_4) that is present in the copper-silver ores. In some crustiform occurrences, it is rhythmically interlayered with bornite (fig. 33) or, in places, with tetrahedrite. It is evidently contem-

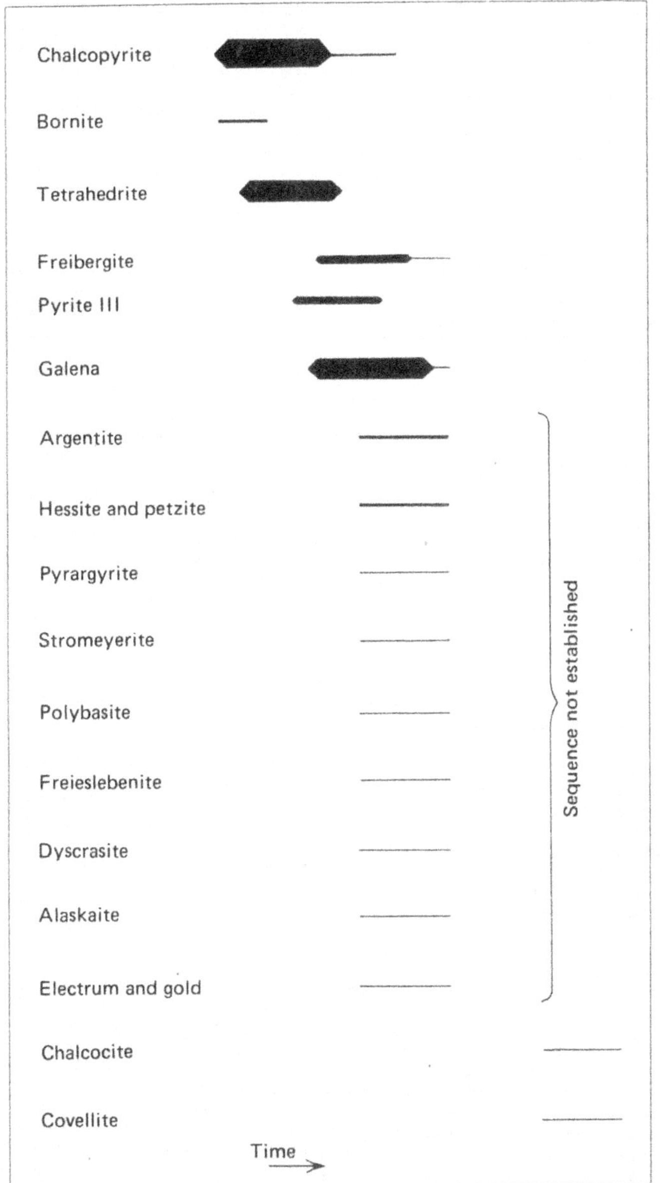

FIGURE 32. — Paragenetic sequence and relative abundance of ore minerals in copper-silver ores in chimney ore bodies, exclusive of host pyrite; widths of bars are proportional to abundance.

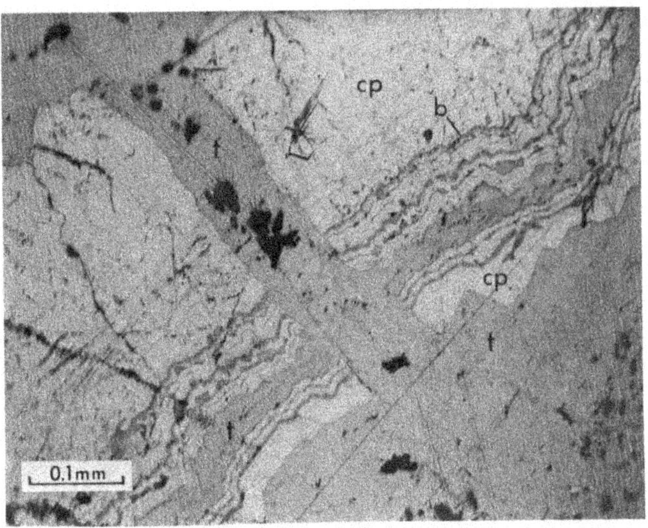

FIGURE 33. — Veinlet of tetrahedrite (t) cutting interbanded chalcopyrite (cp), bornite (b) and tetrahedrite.

poraneous with both bornite and tetrahedrite, though some tetrahedrite is distinctly younger (fig. 33). Some encrusting chalcopyrite is in coarse intergrown crystals (fig. 34), but most of it is finely granular.

All the chalcopyrite is argentiferous, though in a widely varying degree (table 7). Some is also distinctly bismuthian. Gold content of some of the copper-silver ore seems to vary with copper content; from this, we infer that the chalcopyrite is also auriferous, though we have no analytical data for gold in this mineral. A series of samples from the 4-5 chimney (table 7) suggests that the silver content of chalcopyrite decreases irregularly with depth in the ore body.

TETRAHEDRITE AND FREIBERGITE

Tetrahedrite ($(Cu,Fe)_{12}Sb_4S_{13}$), including the silver-bearing variety freibergite ($(Cu,Fe,Ag)_{12}Sb_4S_{13}$), is a major ore mineral and the most abundant of the sulfosalt minerals in the copper-silver ores. Tetrahedrite is generally associated with chalcopyrite (fig. 33), but

TABLE 7. — *Minor element content of chalcopyrite in copper-silver ore, Eagle mine, Gilman district*

[Quantitative spectrographic analyses by J. D. Fletcher, U.S. Geological Survey. Values are in percent. Elements also present in all samples as traces but which are below sensitivity limits are Cd, < 0.02; Ni, < 0.0004]

Sample No. Field	Sample No. Laboratory	Location	Ag	Bi	Ga	Ge	In	Mn	Pb	Sn
11T41	D1452	Lower No. 3 chimney, 15 level.	0.48	0.034	0.001	0.003	0.007	0.011	0.11	0.022
72T40	D1446	18-35 chimney, 17 level	.058	.024	<.001	<.002	.003	.024	.64	.004
		4-5 chimney:								
93T41	D1448	17 level	.18	<.004	<.001	<.002	.004	.010	.009	.019
96T41	D1449	18 level	1.20	<.004	.007	.017	<.002	.014	.007	<.002
108T41	D1450	19 level	.064	.12	.003	<.002	.004	.014	.088	.056
66T40	D1447	20 level	.043	<.004	<.001	<.002	.002	.017	.004	.035

FIGURE 34. — Coarsely crystalline chalcopyrite coated by black finely crystalline freibergite.

some occurs alone as crusts of intergrown crystals in cavities in the host pyrite bodies of the chimneys. Freibergite generally occurs as finally crystalline drusy coatings on other minerals (fig. 34), including tetrahedrite. Some of it is associated with polybasite and dyscrasite, and some is observed in polished section to have been replaced by hessite-bearing galena. Freibergite probably is the largest single source of silver in the copper-silver ores, though many other silver-bearing minerals are present.

PYRITE III

A small amount of pyrite (FeS_2) classed as pyrite III, was deposited after chalcopyrite, bornite, and some tetrahedrite in the copper-silver ores of the chimney ore bodies. Earlier, pyrite had been unstable, for the massive pyrite of the chimneys consisting of pyrites I and II was locally corroded and made spongy before chalcopyrite was deposited. Pyrite III occurs in part as small shiny uncorroded cubic and pyritohedral crystals resting on corroded pyrites I and II and, in places, on chalcopyrite. It occurs also in fine-grained banded veinlets that cut chalcopyrite and also as hollow crystals in late quartz.

We class as pyrite III a coarsely crystalline auriferous pyrite that is present locally in the ore deposits in Sawatch Quartzite and in the veins in Precambrian rocks, though in the absence of other minerals, identity of the pyrite cannot be conclusively established. The pyrite is in cubes and octahedrons in pockets and small veins within the pyrite I bodies in the quartzite. The pyrite of most such pockets and veins exposed by the mine workings was selectively mined, and very little remains to view. Octahedral and cubic pyrite in a vein mined in the Doddridge winze of the Ground Hog mine (pl. 2*A*) contained 21 ounces gold and 39 ounces silver per ton (Means, 1915, p. 17). Octahedral pyrite younger than black sphalerite and associated with resinous sphalerite was observed in places in the veins in Precambrian rocks in the Ben Butler and Star of the West mines (pl. 2*E*). Pyrite was mined in places in these mines and was left standing in others; the mined parts presumably contained auriferous pyrite III.

GALENA

Galena (PbS) is fairly abundant in the copper-silver ores as indicated by the substantial lead production in the period 1935–39 (table 2) when only these ores were mined. It occurs as an inconspicuous interstitial filling in porous pyrite (pyrites I and II) and as crystalline crusts on pyrite and chalcopyrite. In addition to being the chief lead mineral in the ores, much of it is rich in silver (table 6), and it also contains gold. The silver is principally in inclusions of argentite and hessite (fig. 35). Inclusions of freieslebenite and dyscrasite have also been noted. Gold occurs as inclusions in the hessite inclusions and also in the form of rare petzite and electrum inclusions. Not all the galena is equally rich in silver. Though our samples from parts of the chimneys are sparse, available information indicates that galena at the tops and bottoms of the chimneys is poor in inclusions and may be no richer in silver than the galena of the manto ore bodies. Such galena possibly dates from the manto stage of mineralization, but we have no definitive data about this.

Silver-rich galena probably was the chief source of the silver in the oxidized and secondarily enriched sulfide ores in the mines in dolomite rocks at Red Cliff, though no specific source is indicated in the brief description of the ores by Crawford and Gibson (1925, p. 83–86).

ARGENTITE

Silver sulfide (Ag_2S) is here referred to as argentite, though some probably is in the dimorph acanthite, to which argentite may invert with lowering temperature. The chief observed occurrence of argentite is as inclusions in galena (fig. 35). In one polished section, it was observed also in association with freibergite, which it

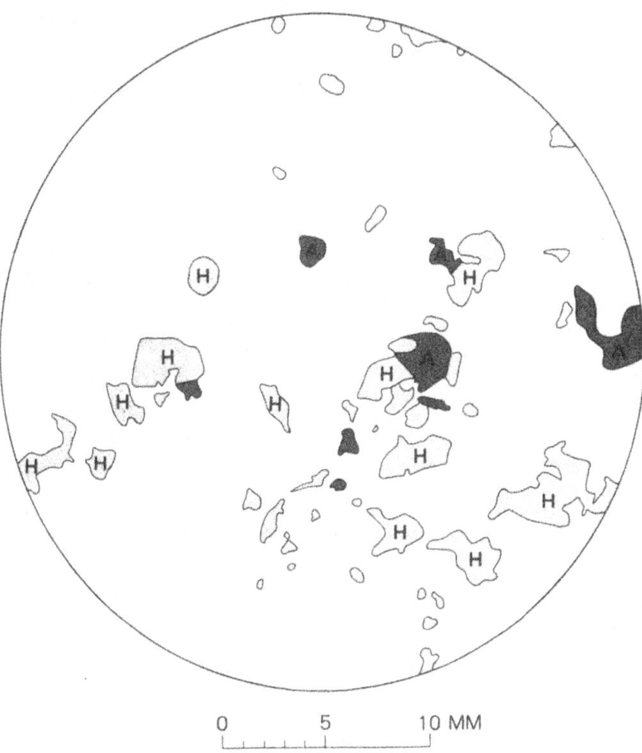

FIGURE 35.—Sketch of polished section of galena in chimney ore, showing inclusions of hessite (H) and argentite (A).

embays and seems to have replaced. Very fine grained argentite may be present in some silver-rich sooty coatings of chalcocite on other sulfide minerals, especially pyrite.

HESSITE AND PETZITE

Hessite (Ag_2Te) and rare petzite (Ag_3AuTe_2) have been observed in the copper-silver ores only as inclusions in galena (fig. 35). They occur in larger quantity as small veinlets in the Sawatch Quartzite. As seen in basal strata of the quartzite in the Star of the West and Ben Butler mines, the veinlets are fillings in random joints, and few of them are continuous for more than 2 or 3 feet ($\approx$75 cm). They range in width from mere films to 0.25 in. (6 mm). Though long referred to as petzite in local usage, they consist principally of hessite that contains microscopic veinlets of petzite and associated free gold (fig. 36).

Though not observed by us, telluride minerals also occur in the ore bodies in the Rocky Point zone of the Sawatch Quartzite. Rich telluride ore was produced from the Ben Butler mine (Olcott, 1887), the main workings of which are in the Rocky Point zone above those shown on plate 2*E*. A pocket of hessite was found in the Forgy incline of the Ground Hog mine. Hessite and sylvanite ($(Ag,Au)Te_2$) are reported to have come from the Champion mine (Eckel, 1961, p. 316). The occurrence of telluride minerals in unspecified mines in the quartzite was noted by Pearce (1890). Telluride minerals may have been an important source of the gold and silver in the oxidized ores in the quartzite.

MINOR ORE MINERALS

Several silver-lead and silver-copper minerals occur in minor quantity in the copper-silver ores of the chimney ore bodies, principally as microscopic veinlets or inclusions in chalcopyrite, tetrahedrite, freibergite, or galena. As these minor ore minerals generally occur singly, their mutual paragenetic relations are uncertain.

Pyrargyrite (Ag_3SbS_3) occurs in scattered localities as veinlets that cut and replace freibergite and chalcopyrite. It is also present, though rare, in ores in the zone of secondary sulfide enrichment in the deposits in Sawatch Quartzite. Stromeyerite ($CuAgS$), a rare mineral in the copper-silver ores, occurs as microscopic veinlets that cut and replace pyrargyrite, freibergite, and chalcopyrite. Polybasite ($(Ag,Cu)_{16}Sb_2S_{11}$) was observed in a few samples as inclusions in freibergite and in a few as veinlets in galena. Freieslebenite ($Pb_3Ag_5Sb_5S_{12}$) was observed in company with argentite and hessite as inclusions in richly argentiferous galena from the lower parts of the No. 3 and 18-35 chimneys. Dyscrasite (Ag_3Sb) occurs in a few localities as a vug filling younger than freibergite. In such occurrences it forms shiny silver-gray aggregates that display a few pyramidal crystal faces. In one sample from the lower part of the 4-5 chimney, dyscrasite is intergrown with alaskaite ($Pb(Ag,Cu)_2Bi_4S_8$) (fig. 37). Dyscrasite has also been observed as blebs and veinlets in galena and as replacement grains in freibergite. Electrum (Au,Ag)

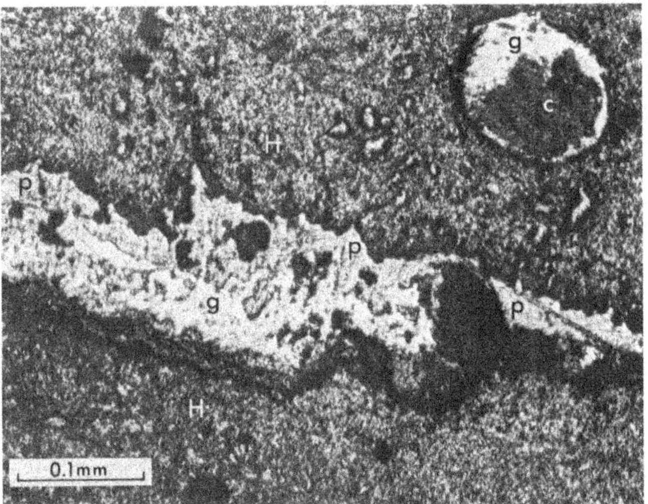

FIGURE 36.—Hessite (H) containing inclusions of petzite (p), gold (g), and a carbonate mineral (c). Specimen from veinlet in quartzite, Star of the West mine.

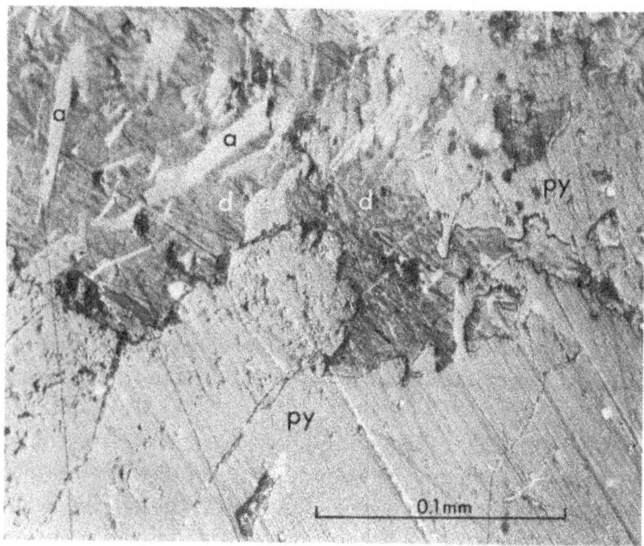

FIGURE 37. — Intergrowth of alaskaite (a) and dyscrasite (d) filling cavities in corroded pyrite (py).

was observed in only two samples. In one, it is a sparse inclusion in galena; in the other, it forms veinlets that cut tetrahedrite. Ferberite ($FeWO_4$) was seen in one sample of ore from the 4-5 chimney, where it occurs as much corroded crystals enclosed in chalcopyrite. Though of no economic significance in the copper-silver ores, the ferberite indicates that tungsten was a minor component of the suite of elements characteristic of these ores.

Chalcocite (Cu_2S) is a widely distributed but volumetrically minor component of the copper-silver ores. It occurs principally as a sooty coating on other minerals, notably chalcopyrite and pyrite, and also as thin replacement rinds on chalcopyrite and as tiny replacement veinlets in chalcopyrite and tetrahedrite. Covellite (CuS) is also present in very small quantities as replacement veinlets in chalcopyrite. In the ore deposits in quartzite, chalcocite is an important ore mineral in the zone of secondary enrichment where it occurs principally as a coating on pyrite. Though not abundant, such chalcocite and associated sooty argentite made minable parts of the otherwise nearly barren pyrite bodies. We infer that the chalcocite and covellite in the chimney ore bodies formed by enrichment processes resulting from oxidation of the upper parts of the ore bodies, just as in deposits in quartzite.

Other copper-silver minerals, not seen by us, are reported by Radabaugh, Merchant, and Brown (1968, p. 653). These are the bismuthian minerals matildite ($AgBiS_2$) and beegerite ($Pb_6Bi_2S_9$), and the arsenical minerals proustite (Ag_3AsS_3), pearceite ($(Ag,Cu)_{16}As_2S_{11}$), and tennantite ($(Cu,Fe)_{12}As_4S_{13}$). Though probably present, the three arsenical minerals are much subordinate to their antimonial counterparts — pyrargyrite, polybasite, and tetrahedrite. Stephanite (Ag_5SbS_4) and tetradymite (Bi_2Te_2S) are reported by Titley (1968).

OXIDIZED ORES

Four general kinds of oxidized ores have been mined in the Gilman district: (1) lead-silver ores in the oxidized parts of ore bodies in dolomite rocks, (2) gold-silver ores in the oxidized parts of ore bodies in quartzite, (3) zinc carbonate ores, principally in dolomite rocks, and (4) manganese-iron oxide ores, mined only from deposits in dolomite rocks but widespread also in the quartzite. Principal ore minerals in the lead-silver ores are anglesite, cerussite, and cerargyrite; in the gold-silver ores, native gold and cerargyrite; in the zinc carbonate ores, smithsonite and hemimorphite; and in the manganese-iron oxide ores, psilomelane, pyrolusite, goethite, and limonite. Common to all the kinds of ores are various sulfate minerals, principally of iron. Except in the quartzite, most of the workings in the oxidized zone were inaccessible to us; therefore, we rely heavily on early descriptions in the accounts that follow.

LEAD MINERALS

According to Emmons (1887), the lead mineral in the oxidized lead-silver ore at Gilman is anglesite ($PbSO_4$), the presence of which he attributed to incomplete oxidation. From the context of his discussion, it is inferred that his observations were made principally in the lower part of the oxidized zone in the Black Iron mine. Except in the lower parts of the oxidized zone, cerussite ($PbCO_3$) probably is the chief lead mineral, as indicated by Crawford and Gibson (1925, p. 50). Both minerals occur as "sand carbonates" and as "hard carbonates" in the parlance of the Leadville district (Emmons and others, 1927, p. 227–229). Sand carbonates are loose to friable aggregates of crystals of coarse-sand size; they may be white or stained to varying degree by iron and manganese oxides. Hard carbonates consist of crystals of cerussite or anglesite embedded in a matrix of iron oxides or a mixture of iron oxides and cryptocrystalline silica. Both varieties generally contain admixed cerargyrite or other silver halide minerals.

SILVER MINERALS AND GOLD

The principal silver mineral in both the lead-silver ores in dolomite rocks and the silver-gold ores in quartzite is cerargyrite (AgCl), also known as horn silver. In the deposits in dolomite rocks, the mineral is generally intermixed with cerussite or anglesite, though in rich silver ore in the Horn Silver mine at Red Cliff it occurs in a matrix of iron and manganese oxides (Crawford and Gibson, 1925, p. 86). In the deposits in quartzite, mixtures of cerargyrite and gold occur as nodules or disseminated grains either in iron oxides or in a clayey material known during the early days of mining in the district as "talc ore." This material, of the color and consistency of putty, consists of clays, aluminous sulfate minerals, and quartz that is residual from the quartzite. "Nuggets" in this material are aggregates of gold particles in a matrix of cerargyrite and are generally concentrated in depressions in the floor of hard quartzite underlying the clayey seam (Guiterman, 1891). Silver also occurs in some places as native or wire silver (Crawford and Gibson, 1925, p. 86) and as the material known to the early miners as "black sulphurettes," which presumably is a powdery form of argentite (acanthite).

ZINC MINERALS

Zinc carbonate ore consists mainly of smithsonite ($ZnCO_3$) but contains some hemimorphite or calamine ($Zn_4Si_2O_7(OH)_2 \cdot H_2O$) and hydrozincite ($Zn_5(OH)_6(CO_3)_2$). The ore occurs in pockets on the undersides of the oxidized parts of the manto ore bodies, where it replaced dolomite and siderite (Heyl, 1964, p. 57; Crawford and Gibson, 1925, p. 51). Much of it is heavily impregnated with iron and manganese oxides. Red-brown ferruginous zinc carbonate ore occurs also in small bodies in the Sawatch Quartzite, where it was evidently deposited by waters descending from the manto ore bodies. A body of this ore in the Tip Top mine was worked as early as 1907, but the main production of zinc carbonate ore was not made until after such ore was recognized in quantity in old workings in the Leadville district in 1910 (Loughlin, 1918).

Goslarite ($ZnSO_4 \cdot 7\ H_2O$) is fairly abundant on the walls of old zinc stopes and in joints in dolomite rock below such stopes. It forms dense efflorescent crusts of long needles, frequently in the form of satiny white beards (fig. 38). Much of the goslarite is magnesian and grades to epsomite ($MgSO_4 \cdot 7H_2O$) of similar appearance. The goslarite quickly dehydrates to a dull white powder when removed from the mine workings. It is not significant as an ore mineral.

MANGANESE AND IRON OXIDES

Mixtures of manganese and iron oxides characterize the outcrops and oxidized upper parts of the manto ore bodies in dolomite rocks and the lower parts of the oxidized zone in the ore bodies in quartzite. Manganese oxides are relatively abundant in the deposits in dolomite rocks and were

FIGURE 38. — Goslarite (hydrous zinc sulfate) beards on zinc sulfide ore. Similar beards on dolomite near ore are epsomite (hydrous magnesium sulfate) or intermediates between goslarite and epsomite. Long beards at left and right are about 8 in. (20 cm) in length.

mined in places either as metallurgical or fluxing ore (Umpleby, 1917). In the quartzite, manganese oxides are far subordinate to iron oxides but are locally concentrated in pockets or lenses in the iron oxides. In both the dolomite and the quartzite settings, the manganese was derived from manganiferous siderite, and the iron was derived from siderite and pyrite. Where relatively pure, the manganese oxide consists principally of hard massive material generally called psilomelane ($BaMn^{+2}Mn_8^{+4}O_{16}(OH)_4$), though it may contain other manganese mineral species. It was not studied in detail. Sooty manganese oxide assumed to be pyrolusite (MnO_2) coats openings and forms small veinlets in the psilomelane. A few vugs are lined with tiny pyramidal crystals of braunite ($3(Mn,Fe)_2O_3 \cdot MnSiO_3$).

The iron oxides are classed collectively as limonite. They consist in part of goethite ($HFeO_2$) and in part of unidentified hydrated ferrous oxides. As judged by the predominant brown to ocher color, anhydrous ferric oxides are sparse. Iron sulfates are generally present in the limonitic material and are abundant in soft, plastic varieties. In the ore bodies in the quartzite, the limonitic material in places is stratified ocher (fig. 39). The ocher consists principally of goethite but contains kaolinite, quartz, hematite, manganese oxides, and iron sulfates.

SULFATE MINERALS

Dark-yellow jarosite ($KFe_3(SO_4)_2(OH)_6$) is an abundant constituent of the limonitic and ocherous materials in the oxidized ore bodies where it is intimately intermixed with the iron oxides. It occurs

FIGURE 39. — Stratified ocher in oxidized zone in Rocky Point mine. Fold and fault structure is caused by gravity sliding in the plane of the ore bed, which dips about 10° to left. The ocher is soft, plastic, and light yellow to dark brown. Note hammer for scale.

also in nearly pure form as nodules in the puttylike clay that contains gold and cerargyrite in the oxidized deposits in quartzite. Greenish-yellow copiapite ($Fe^{+2}Fe_4{}^{+3}(SO_4)_6(OH)_2 \cdot 20H_2O$) coats surfaces of limonitic materials and occurs also as an impregnation in clayey materials. Dull-green coquimbite ($Fe_2(SO_4)_3 \cdot 9H_2O$) occurs in places as a coating on oxidized ore and wallrocks in old mine workings. Blue-green to yellowish-gray melanterite ($FeSO_4 \cdot 7H_2O$) is abundant as efflorescent crusts in old mine workings, particularly in the zone of transition from oxidized to sulfide ores. Pinkish-brown roemerite ($Fe^{+2}Fe_4^{+3}(SO_4)_4 \cdot 14H_2O$) occurs in small granular aggregates in black argillaceous material at the base of the oxidized zone in dolomite rocks where it is associated with melanterite, gypsum, and partly oxidized pyrite. Granular brown szomolnokite ($FeSO_4 \cdot H_2O$) occurs sparsely in the same setting as roemerite. Fibroferrite ($Fe(SO_4)(OH) \cdot 5H_2O$) is also reported (Eckel, 1961, p. 146). Alunite ($KAl_3(SO_4)_2(OH)_6$) is a common constituent of the puttylike clayey materials in the intensely oxidized upper parts of the ore bodies in quartzite. In jarosite nodules in this material, many of the individual jarosite grains have a nucleus of alunite. Gypsum ($CaSO_4 \cdot 2H_2O$) is a common constituent of the oxidized ores in dolomite rocks, where it occurs either as fine-grained interstitial matter in aggregates of the ore minerals or as small satiny crystals of the variety selenite. Blue to greenish-blue chalcanthite ($CuSO_4 \cdot 5H_2O$) occurs in places as films on the walls of workings in quartzite and, also, as small stalactites along joints exposed by these workings. It is nowhere abundant enough to be significant as an ore mineral.

ORE DEPOSITS IN DOLOMITE ROCKS

The ore deposits in dolomite rocks — by far the most productive in the Gilman district — comprise the manto and chimney ore bodies at Gilman (fig. 11) and the scattered and irregular small ore bodies in the mines at Red Cliff. The ore bodies of all three kinds are fundamentally replacement deposits, though substantial fractions of the ore and gangue minerals in each were deposited in open spaces in the generally porous ore bodies. Contacts between the sulfide ore bodies and the dolomite wallrocks are sharp. Ore and gangue minerals typically constitute 100 percent of the material on one side of a knife-edge contact, and dolomite rock that contains metals only in the parts-per-million range constitutes the other side.

MANTO ORE BODIES

The manto ore bodies, the largest and most productive of the ore bodies in dolomite rocks, are in the Leadville Dolomite, principally in the upper part (fig. 25). Four main mantos extend generally northeastward in the plane of the strata — which dip about 12° — to a northwest-trending line of chimney ore bodies and short connecting strike mantos (fig. 11). The main mantos are irregular pipelike bodies 2,500–4,000 ft (762–1,220 m) long. Except at their upper ends, they are generally elliptical in cross section and are 50–400 ft (15–122 m) wide and 2–150 ft (0.6–46 m) thick. They flatten and widen in the updip direction to blanketlike bodies that were the source of the term manto. Except where oxidized, the mantos consist principally of black sphalerite (marmatite), pyrite, siderite, and subordinate galena. Typical manto ore is a mixture of these minerals in various proportions. Siderite occurs additionally as a nearly monomineralic shell around the manto bodies in many places. At their lower ends, near their junctions with the chimneys, the mantos are zoned in cross section. An inner core consists largely of pyrite but contains subcommercial amounts of sphalerite and a little siderite. This pyrite core is continuous with the pyrite core of the connected chimney (fig. 25). The pyrite core of the manto is surrounded by a nearly monomineralic shell of black sphalerite, which is continuous with that of the sphalerite shell of the chimney and of the strike mantos between chimneys. An outer shell that consists principally of siderite is continuous with the siderite shell of the chimney.

The grade of sulfide ore in the mantos varies with locality. Radabaugh, Merchant, and Brown (1968, p. 659) stated that the average grade is approximately 12.0 percent zinc, 2.0 percent lead, 1.0 ounce silver, and 0.02 ounce gold per ton as determined from mine production over a long term and that the grade mined through yearly periods has ranged from 11.7 to 17.0 percent zinc and 1.2 to 2.4 percent lead.

The oxidized upper parts of the mantos have not been worked in recent decades and were largely inaccessible to us. These parts, which extend to distances of about 1,000 ft (305 m) downdip from the outcrop, were mined principally for the silver in a lead-carbonate or lead-sulfate ore. Though very rich pockets are reported, most ore of this type contained 4 to 25 ounces silver and a trace to 0.5 ounce gold per ton and 8 to 15 percent lead (Olcott, 1887; Radabaugh and others, 1968, p. 659). Oxidized zinc ore is not abundant, but 6,000 to 10,000 tons of unspecified grade was produced in the period 1912–17 (Heyl, 1964, p. 57). Abundant manganese oxide in the oxidized zone suggests that siderite may have been more abundant in the upper parts of the mantos than elsewhere. Umpleby (1917) reported that 200,000 tons of fluxing ore averaging 15 percent manganese, 38 percent iron, and 1–2 percent silica was mined from the Black Iron ore body (No. 2 manto) prior to 1917; he estimated that 750,000 tons of such material remained in that body and the Iron Mask ore body (No. 1 manto).

CHIMNEY ORE BODIES

The chimney ore bodies are subvertical funnel-shaped bodies that cut across the strata at a high angle to the bedding; their axes plunge about 70° NE. (fig. 25). The 4-5 chimney, the largest, is about 400 ft (122 m) in diameter at the top. The main chimneys (fig. 11) extend through the full thickness of the Leadville Dolomite, the Gilman Sandstone (which is in part dolomite), and the Dyer Dolomite. These three units total about 240 ft (73 m) in thickness in the vicinity of the chimneys. As replacement deposits, the chimneys terminate downward at the top of the Parting Formation (table 1). Below that level, sulfide veinlets in joints in the quartzite of the Parting make small areas minable as deep as the Harding Sandstone.

The chimneys are compound ore bodies that consist in part of the pyrite-sphalerite-siderite assemblage of the mantos, with which they are continuous (fig. 25), and in part of the younger assemblage of copper, silver, lead, and gold minerals that constitute copper-silver ore. The older, or mantolike, assemblage is distributed zonally. Pyrite is largely concentrated in central cores of the chimneys; sphalerite is principally in discontinuous shells surrounding the pyrite cores, and siderite is chiefly in discontinuous outer shells. Boundaries between these units are generally gradational but are sharp in places.

The copper, silver, lead, and gold minerals impregnate parts of the pyrite cores of the chimneys, transforming those parts to copper-silver ore (fig. 25). The parts of the cores that were not enriched by these minerals remain as pyrite that contains only subcommercial amounts of black sphalerite (marmatite). The various minerals of the copper-silver ores are erratically distributed and as a consequence the grades of the ores vary widely. Radabaugh, Merchant, and Brown (1968, p. 659) reported an average for all copper-silver ore mined of 18.7 ounces silver and 0.08 ounce gold per ton and 3.5 percent copper but noted that some of the ore contained several thousand ounces of silver and several tens of ounces of gold per ton. The distribution of rich ore within and among the chimneys is not known to us. The chimney ore bodies consitute very heavy ground, which requires large amounts of timber for support. Thus, at any given time they are exposed in only a few square-set working faces 5 by 6 ft (1.5 by 1.8 m) in dimensions. Mineralogic mapping would require daily mapping through periods of years, an activity not feasible for us. Such stope assay maps as we have seen indicate abrupt large changes in the contents of the valuable metals and only a very general correlation between pairs of metals. On the basis of about 50 samples from the 4-5 chimney, we see some indication that silver is mainly in galena in the upper part of the chimney and mainly in freibergite and other sulfosalt minerals in the lower part. We have no information on the mineralogic character of the very rich gold ore beyond the fact that if gold is high, copper is high, though the reverse does not apply.

ORE DEPOSITS AT RED CLIFF

The ore deposits at Red Cliff have been worked in several small mines, the best known of which are the Horn Silver, Liberty, Wyoming Valley, and Silurian mines (pl. 1). The deposits are all in the Leadville Dolomite, at or near the top. The dolomite in the vicinity of the deposits is mainly zebra rock and is strongly sanded in many places. As described by Crawford and Gibson (1925, p. 83–86), the ore bodies are small, irregular replacement deposits along minor faults in brecciated dolomite rock. Almost all the ore mined was either argentiferous lead carbonate ore or partly oxidized and secondarily enriched pyrite-sphalerite-galena ore. The ores were valuable principally for silver; many lots of the ores contained hundreds of ounces of silver per ton. Some pockets of silver ore in the Liberty mine are reported to have contained as much as 0.77 ounce gold per ton. A 2-in. (5-cm) streak of bismuth ore is reported from a small mine near the Horn Silver (Mining and Scientific Press, 1910, v. 101, p. 756). "Graphite" mentioned in the same news note probably is highly carbonaceous shale from the Belden Formation. Such shale is seen in many places to have a graphitic luster on slip surfaces.

The high silver content of the Red Cliff ores and the presence — locally at least — of gold and bismuth suggest affiliation with the copper-silver ores of the chimney ore bodies at Gilman. The Red Cliff area is, however, a separate area of mineralization, almost a mile from the nearest chimney ore body at Gilman.

STRUCTURAL VARIETIES OF ORES

The ores in the manto and chimney ore bodies differ in structure from place to place depending on the character of the dolomite rock that was replaced by ore and gangue minerals. Three main structural varieties of ore are recognized. These varities have no relation to ore composition but are significant to understanding of mineralization processes. They also influence mining and ore treatment costs, for they differ markedly in strength as related to mine openings and in ease of crushing in milling operations.

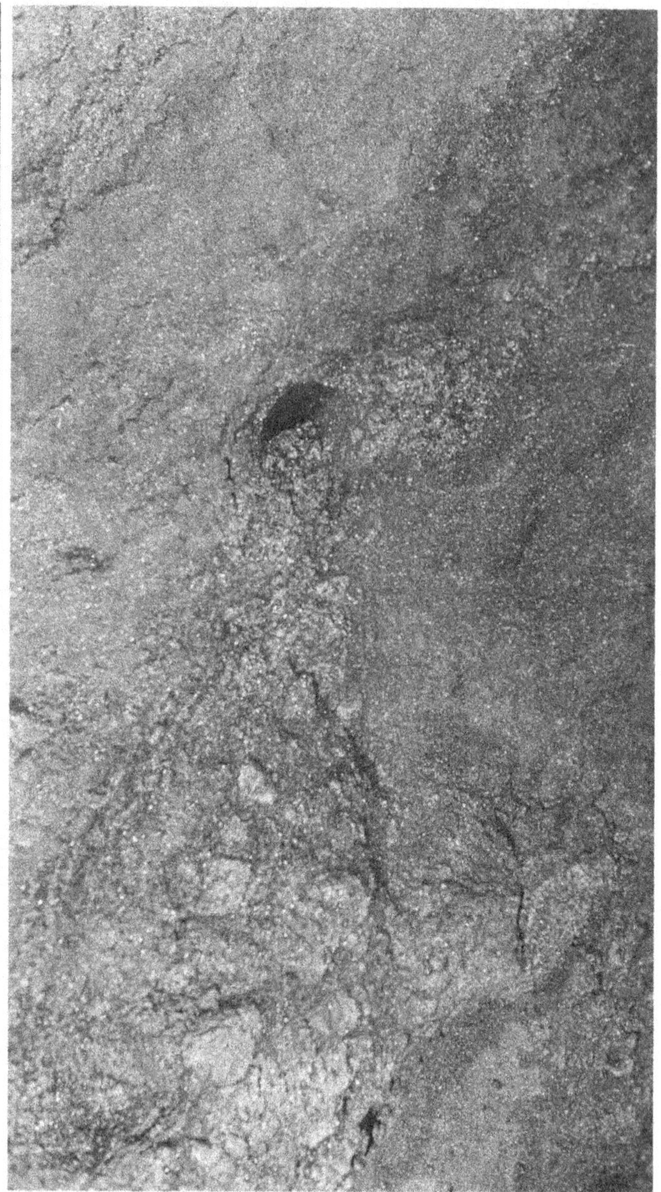

FIGURE 40. — Sand and rubble ore. Dolomite cave sand and rubble replaced by pyrite and sphalerite, No. 3 manto, 17 level, Eagle mine. White specks are reflections from pyrite crystals. Area of photo about 1.5 × 3 ft (0.45 × 0.9 m).

One variety, which we term bedrock ore, is a hard, tough, compact intergrowth of ore and gangue minerals in which are preserved many features of the precursor dolomite strata, such as bedding, zebra structure, chert lenses, shaly partings, and joints. This variety resulted from replacement of bedded dolomite rock. A second variety, termed sand ore, has a sandy texture and is generally loosely granular, though locally firmly bonded (fig. 40). Preserved in it are sedimentary structures identical with those observed in the dolomite sand deposits in solution channels and caves, described in the section "Rock Alteration." Stratification is evident in many places (fig. 41). This variety of ore resulted

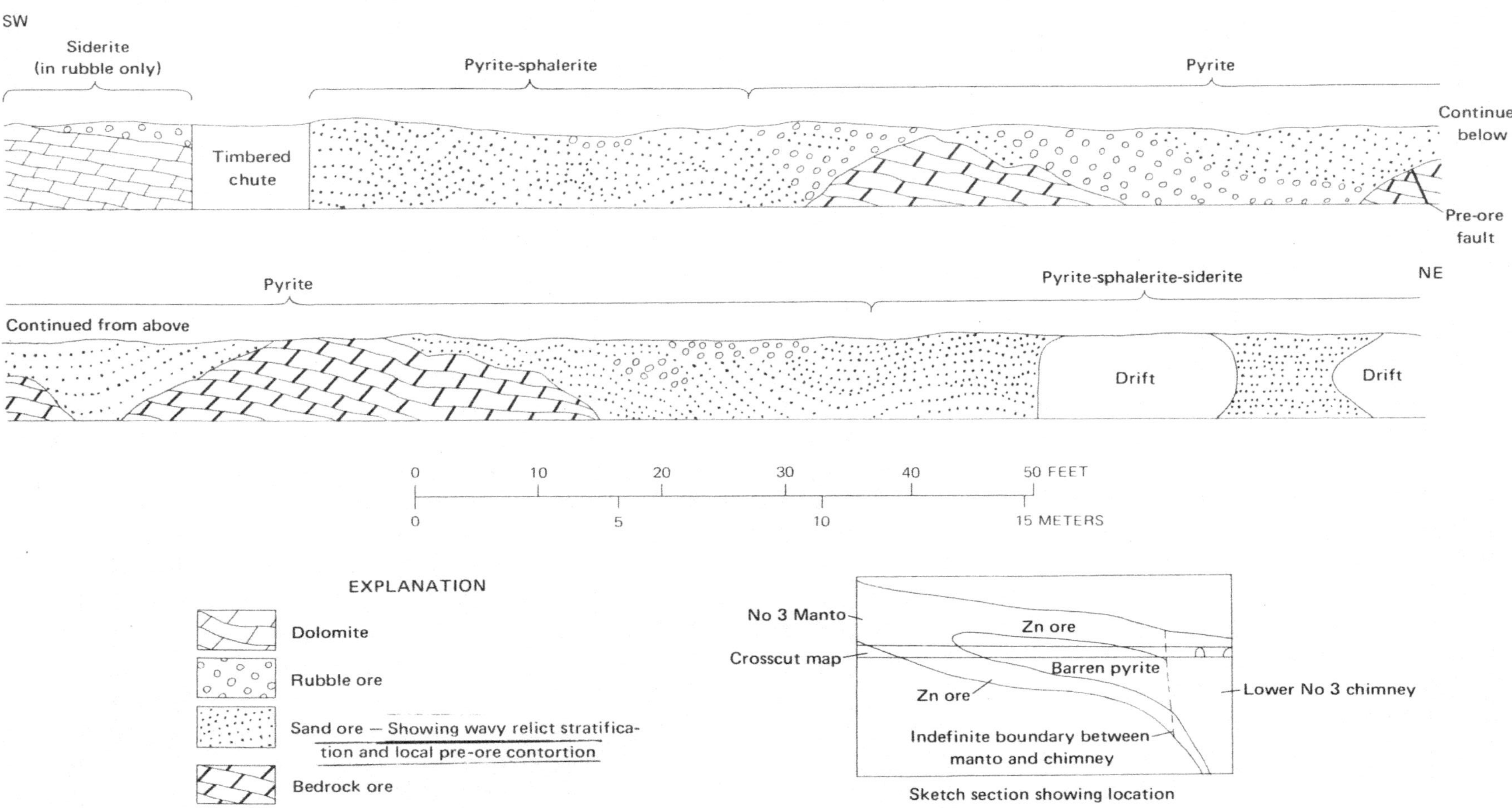

FIGURE 41. — Wall of crosscut near junction of No. 3 manto and Lower No. 3 chimney, 17 level, Eagle mine, showing distribution of ore types.

from replacement of dolomite sand deposits. The grain size, however, is coarser than in the original sand. A third variety of ore, termed "rubble ore," has a lumpy structure similar to that of cave rubble (fig. 42), and it resulted from replacement of such rubble. Just as the cave rubble, rubble ore contains in places foreign materials, such as fragments of porphyry, black shale, and chert. The three structural varieties of ore are present together in many parts of the ore bodies (fig. 43). They also intergrade. Bedrock ore that formed by replacement of strongly fractured dolomite or crackle breccia grades insensibly to rubble ore. Bedrock ore that replaced strongly sanded dolomite grades to sand ore.

Additional types of breccia or rubble formed during and after mineralization. Taluslike ore breccia present in some places (fig. 44) consists of fragments of sulfide ore bonded by sulfides. Such breccia evidently formed by collapse of sulfide ore into cavities during the mineralizing process. Cave rubble that contains fragments of early ore minerals, such as siderite and pyrite, and is coated by paragenetically younger sphalerite (fig. 19) indicates that rubble continued to accumulate during mineralization. Bodies of collapse breccia consisting of rocks from the Belden Formation and the porphyry sill in the Belden are present in the upper parts of the manto and chimney ore bodies in many places (fig. 25). This breccia contains fragments of ore, indicating that collapse of the "cap rocks" occurred during or after the mineralizing event. The breccia also indicates the widespread occurrence of cavities in the upper parts of the ore bodies by the time mineralization was drawing to a close or had ended. Though the cavities were probably formed principally by dissolution of dolomite rock, they might also reflect shrinkage or volume changes resulting from the replacement of dolomite by the much denser sulfides. The occurrence of isolated bodies of the collapse breccia as deep stratigraphically as the Dyer Dolomite suggests a redistribution of collapse materials by circulating waters.

FIGURE 42. — Rubble ore, crudely stratified; No. 4 manto, 16 level, Eagle mine. Area of photo about 4 × 7 ft (1.2 × 2.1 m).

STRUCTURE OF MANTOS AND CHIMNEYS

Though it was not possible to observe all parts of the mantos at the time of our studies, an irregular and crooked body of rubble and sand ore apparently extends through the length of each manto. Thus, the mantos evidently coincide with systems of pre-ore channelways and caves formed by dissolution of dolomite rock. The channelways of these systems are very irregular. Rubble and sand ores constitute almost the entire cross sections of the mantos in some places, especially near the chimneys; they constrict to small cross-sectional areas in other places and in the updip parts of the mantos. The rubble and sand ores may lie in any part of the cross section of a manto but most commonly are at the top or along one side. They are commonly underlain and bordered on one or both sides by bedrock ore, though they rest directly on dolomite in places (fig. 43). Rubble ore predominates over sand ore in the upper (southwest or updip) parts of the mantos, and the proportion of sand ore increases downdip toward the chimneys. Bodies of collapse breccia lie within or above the mantos in places, and small bodies of karst silt are present locally (fig. 25). The dolomite wallrocks sag slightly toward the mantos throughout their length and locally sag markedly. The sagging evidently reflects volume losses due to dissolution (sanding).

The chimney ore bodies consist principally of sand and rubble ores in their upper parts — in the Leadville Dolomite — and of bedrock ore in their lower parts — in the Dyer Dolomite. Sand ore is abundant on the foot-

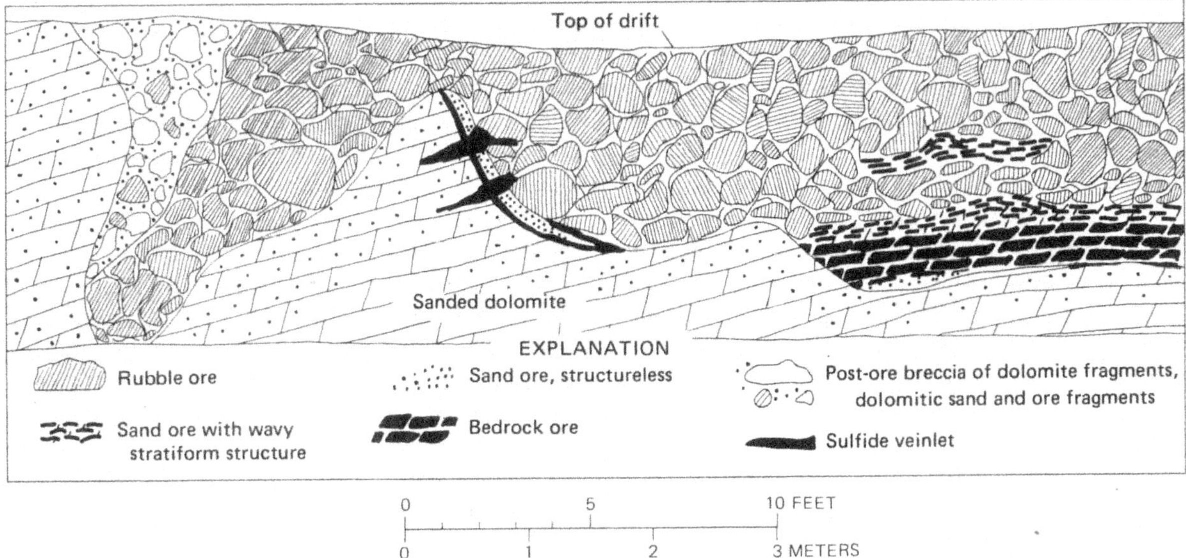

FIGURE 43. — Distribution of rubble, sand, and bedrock ores and post-ore breccia at a locality in No. 3 manto, 15 level, Eagle mine.

FIGURE 44. — Sulfide ore collapse breccia, No. 1 manto, 15 level, Eagle mine. Though open-textured, this breccia is firmly bonded and requires blasting.

wall side, near the junction of chimney and manto (fig. 41). In places, relict stratification in the sand ore dips steeply (20°–30°) into the chimneys. A shell of bedrock ore underlies the sand ore on the footwall sides of the upper parts of the chimneys. Rubble ore is abundant in the upper central parts of the chimneys and along the hanging wall sides. Bedrock ore on the hanging wall sides replaced strongly fractured dolomite strata and hence grades into rubble ore. Numerous bodies of collapse breccia extend into the upper parts of the chimneys, and bodies of karst silt are more abundant in the chimneys than in the mantos (fig. 25). As shown by the Gilman Sandstone, which was not entirely replaced, wallrock strata sag into the chimneys on both the updip and downdip sides, particularly in the upper halves of the chimneys (fig. 25).

GEOLOGIC CONTROLS OF MANTOS AND CHIMNEYS

The manto and chimney ore bodies in dolomite rocks were evidently controlled in location by solution-channel systems of two ages and origins — an older karst system and a younger hydrothermal system. Controls of the karst system have not been identified but may have been joints that existed as early as Early Pennsylvanian time. Controls of the hydrothermal solution-channel system were the karst system and small-scale tectonic features. The dolomite rocks themselves provided a receptive environment for mineralization, as they were more reactive to the rock-altering and mineralizing solutions than any of the other rocks in the district.

As noted in the sections "Stratigraphy" and "Rock Alteration," solution channels of the karst system are identified by the presence in them of karst silt, a distinctive clayey silt that contains chert fragments and chert nodules of distinctive shape. The karst silt is identical with, and in many places continuous with, the discontinuous thin stratigraphic unit known as the Molas Formation, which regionally underlies the Middle and Lower Pennsylvanian Belden Formation. The karst silt, therefore, is pre-Belden in age.

At the time of karst erosion, the Gilman Sandstone was at shallow depth and probably constituted an aquifer in the ground-water circulation system. Dissolution features in the Gilman, described under "Rock Alteration," are inferred to have had karst progenitors, though they were later greatly accentuated by hydrothermal dissolution.

As karst silt is much more abundant in and near ore bodies than elsewhere and as dissolution features in the Gilman Sandstone are largely restricted to the mineralized area, we conclude that karst solution channels were principal elements in the "plumbing system" that provided access for the rock-altering and mineralizing hydrothermal solutions. The full extent of the karst solution channels cannot be known because of the effects of the later hydrothermal dissolution process. Karst channels are identified with certainty only where they contain karst silt. But some may have not contained silt, or they may have had their silt flushed from them. They would then be indistinguishable from the channelways formed by hydrothermal dissolution. The occurrence of karst channelways seems to correlate inversely with thickness of the Leadville in the following manner: Where the Leadville was eroded most deeply in pre-Belden time, karst channelways are present and most numerous; where the Leadville was eroded less deeply, channelways are absent or few. In turn, ore deposits at Gilman are in areas where the Leadville is relatively thin (Radabaugh and others, 1968, p. 662), as is true also in the Leadville district (Tweto, 1968a, p. 696). Though the thinning of the Leadville results largely from the erosional removal of upper strata, it may also reflect the thinning of some surviving strata by karst dissolution. Porosity of upper strata of the Leadville increased in the process.

Channelways created or extended and integrated during the stage of hydrothermal dissolution of dolomite rock are far more numerous and more extensive than those of proved karst origin. They are distinguished from the karst channels by the presence in them of materials younger than the karst silt, namely, (1) rocks from the Belden Formation, (2) altered facies of Leadville Dolomite (zebra rock and pearly dolomite), and (3) most significantly, fragments of porphyry dated as 70 m.y. (Late Cretaceous) in age. As indicated under the two preceding headings, such channelways are preserved as rubble and sand ores in the replacement ore bodies. They extend through all or most of the main ore bodies, and they were clearly the immediate geologic control of the ore bodies.

Some hydrothermal dissolution channels are inferred to be enlargements and extensions of old karst channels. Others are in areas in which there is no firm evidence of karst channels. Dissolution that enlarged old channels or formed new ones was controlled by subtle tectonic features, such as joints and small faults, bedding and tongue faults, and slight warps in the strata. In turn, ore bodies were controlled by these same features.

The rocks throughout the Eagle mine are strongly jointed. The directions of the joints and the numbers of joint sets vary from place to place (fig. 45). They also vary vertically. In many places, joints of certain sets are confined to a section of strata between bedding faults. In some areas, particularly near ore bodies, dolomite rock is sanded along certain joint sets and unsanded along others. Some of the sanded joints extend into or along channels oriented in the direction of the joints. Such channels may be open, may be filled with rubble or dolomite sand, or may be rubble and sand ore.

The fact that joints of certain trends are sanded or channeled whereas those of other trends are not indicates that some joints were "open" and permeable at the dissolution stage and that some were tight and impermeable. The fact that the directions of joints and of the sanded joints differ in different localities indicates that local features controlled both the orientations of joints and which joint sets were "open" or tight at the dissolution stage. We conclude that the directions and the permeabilities of joints were controlled by the combination of bedding-fault movements and slight flexures in the strata. Bedding faults of various magnitudes are numerous, and movements in several directions occurred on them, as discussed under "Structure." Movements of plates of rock in the plane of the bedding in slightly flexured strata would produce local fields of tension and compression. Joints oriented at a large angle to the tensional direction would be pulled open, and those that were open and permeable at the dissolution stage were sanded or channeled. This does not necessarily require tectonic movements concurrent with dissolution, but only that the joints remain in the state they were when their wallrocks were last "parked." Some joints were in existence when zebra rock formed, and some probably existed at the stage of karst dissolution.

The relation of a part of the No. 1 manto to fractures and flexures is illustrated in figure 45. As the map shows, the main part of the manto, which here occupies the full thickness of the Leadville Dolomite, is located near the crest of a slight anticlinal flexure. A subsidiary prong, known as the Venus ore body, is in lower strata of the Leadville along the Venus fault, a tongue fault described under "Structure."

In the area of the main manto, joint sets of several orientations are present. The most prominent and persistent set strikes about N. 70° E., parallel to the trend of the manto. Small faults also have this orientation. The sides of the manto are defined in places by this set of fractures, and fractures of the set are preserved in bedrock ore. Where in dolomite, fractures of the set are marked by sanding. The fractures are interpreted as tension fractures in the crestal area of the anticlinal flexure on which the manto is localized. Much of the ore in the manto is rubble and sand ore that defines a solution channel of the same general trend as the manto, the flexure, and the fractures of N. 70° E. trend. The fractures almost certainly controlled the dissolution that produced the channel that localized the ore body.

Other fractures had a role locally. A set of joints with strikes in the northwest quadrant changes progressively in trend from west-northwest in the western part of the mapped area to northwest in the eastern part. In places, the west-northwest-trending joints define the edge of the ore body, and over a longer distance, the edge jogs along intersecting west-northwest- and east-northeast-trending joints. Joint sets of northeast, north-northeast, and north-northwest trends are present in local areas. Though they have no evident direct effect on the manto, their intersections with the more persistent joint sets created areas of intense jointing that fostered dissolution in the main channelway of the manto.

The Venus ore body illustrates the control of dissolution and consequent mineralization by tongue and bedding faults. As shown by the sketch cross section of the Venus fault and ore body (fig. 45), the ore body consists of a channel of rubble ore centered over the line along which the Venus tongue fault "rolls over" into bedding faults. On the northwest side of this line, blanketlike wedges of rubble ore lie along bedding faults, which were evidently sites of dissolution. On the southeast side, bedding faults are absent and the ore body is thick. We infer that in this area, tensile stresses were created as a result of the northeastward movement of the block northwest of the Venus fault at an angle lower than the dip of the bedding. The tensile stresses inferentially produced fractures that fostered dissolution and the formation of rubble, which subsequently was replaced by ore. Collapse breccia of shale and porphyry over this part of the ore body indicates that dissolution extended to the top of the Leadville.

The chimney ore bodies show the same general types of local controls as the mantos. The main chimneys (fig. 11) are alined along a gentle northwest-trending structural terrace that is corrugated by slight northeast-trending cross flexures. Joints and faults of very small displacements are numerous in the wallrocks and are also preserved in bedrock ore within the chimneys. Fractures of a northeast trend are present in almost all areas, whereas fractures of other trends differ in orientation from area to area and also with stratigraphic level. Many of the fractures of northeast trend are sanded or are marked by rubble channels. They are interpreted as tension fractures in the axial direction of the northeast-trending flexures. The northeast and southwest edges of the ore bodies are bordered by fractures (joints) trending in other directions — principally northwest and east-northeast to west-northwest. In places, northeast-trending faults with vertical displacements of as much as 20 ft (6 m) are evident at the borders of the chimneys, but they die out in short distances along strike and in the vertical direction. They probably formed by slumping of blocks of wallrocks into the chimneys during dissolution at the chimney sites.

At the bottoms of the chimneys, quartzite of the Parting Formation is strongly jointed. Orientations of the joints differ from chimney to chimney. Beneath the 4-5 chimney, joints in the Parting trend N. 45° E., N. 60° E., and N. 75° E. Beneath the Lower No. 3 chimney, they trend N. 50° E. and N. 5° E. Beneath the 18-35 chimney, they trend N. 80° E. to N. 80° W. and N. 20° E. Beneath the Lower No. 2 chimney, they trend N. 50° E. and east-west. The quartzite walls of one or two joint sets beneath each chimney are mildly sericitized, whereas the walls of other sets are fresh and glassy. Veinlets of pyrite occupy joints of all orientations, and quartzite in the upper part of the Parting generally contains disseminated pyrite. Despite the mildness of alteration and mineralization along them, the joints in the Parting Formation (30 ft (9 m) thick) evidently formed a connection in the circulation system between major bedding-fault conduits below and the chimneys above.

A major bedding fault, characterized by a thick seam of clayey gouge, lies at the base of the Parting Formation, in upper strata of the Harding Sandstone. As noted under "Structure," strata above the fault moved updip, to the west or southwest. This movement would produce in the hanging wall a strain shadow on the updip, or leeward side, of any protuberance in the footwall. The gentle flexures evident on the structural terrace that contains the chimneys are believed to constitute protuberances responsible for ovate areas of tension in the strata overlying the bedding fault. At these bumps in the footwall, the gouge of the bedding fault probably thinned or was absent, thus, allowing solutions that moved updip under the gouge to leak through the fault and into the fractures above. These solutions had moved only through granite and quartzite prior to leaking up through the fault, and they were far from equilibrium with dolomite when they entered the carbonate sequence. As they circulated through the fracture system, they extensively dissolved the dolomite beneath the relatively impervious shale of the Belden Formation leaving porous rocks in funnel-shaped bodies. Mineralization began while dissolution was in progress, and the chimney ore bodies were cre-

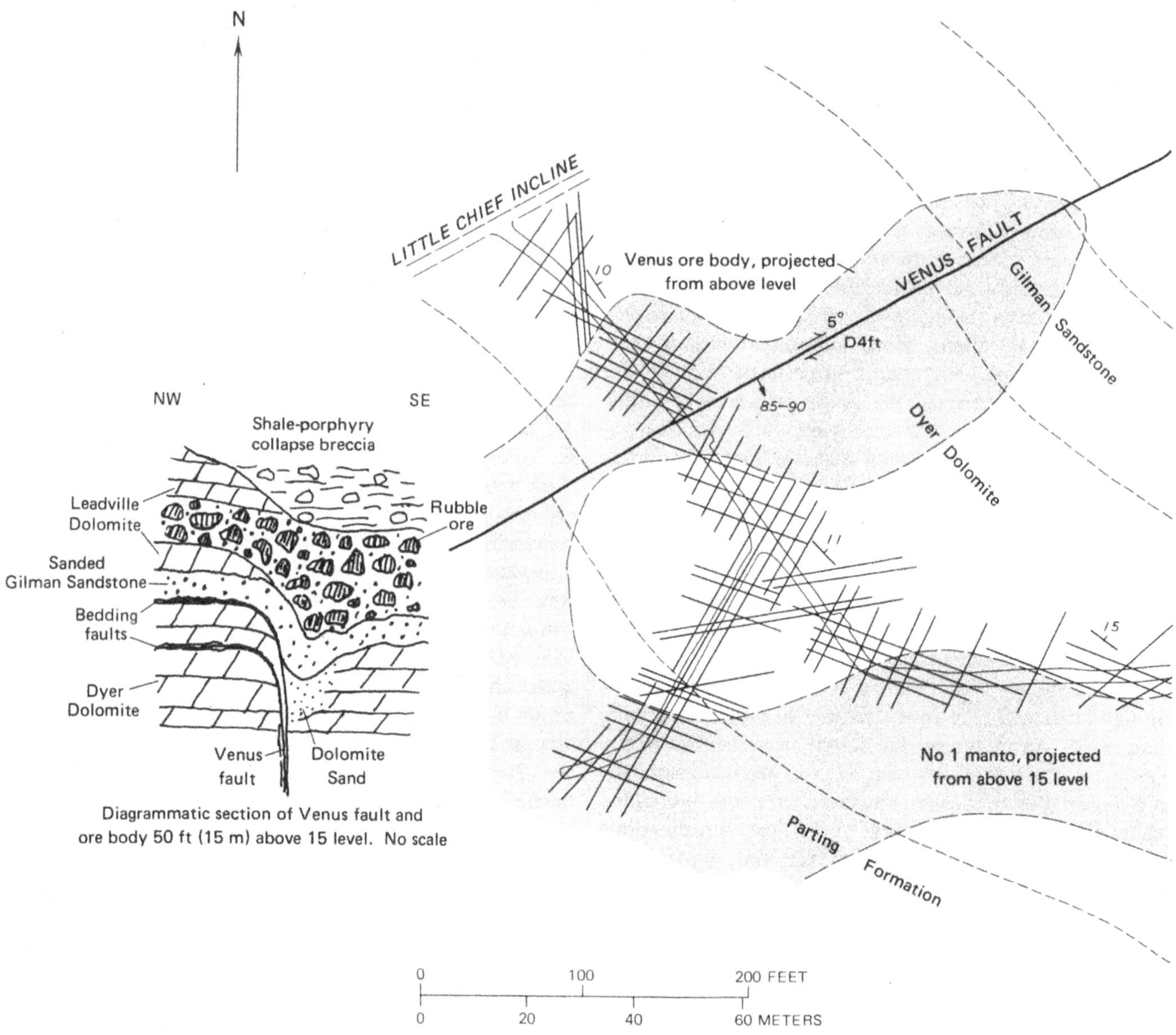

FIGURE 45 (above and facing page). — Part of 15 level in area of No. 1 manto, Eagle mine, showing relation of ore body to fracture system and flexures in strata.

ated at the sites of the porous bodies of dolomite rock.

The route by which the mineralizing fluids reached sites beneath the chimneys is discussed further in the section "Hydrothermal Processes."

ORE DEPOSITS IN SAWATCH QUARTZITE

GENERAL FEATURES

Ore deposits in the Sawatch Quartzite supplied most of the gold and a large part of the silver produced in the Gilman district in the period from 1884 to about 1925 (table 2). Almost all the deposits are in the upper 30–40 ft (9–12 m) of the quartzite in what is known as the Rocky Point zone. Very small silver-gold telluride deposits occur in the basal 30–40 ft (9–12 m) of the Sawatch. Except in the Bleak House mine, production from the deposits in the Rocky Point zone has come principally from the oxidized zone. None of the mines in the quartzite, except the Bleak House, has been worked in recent decades, though several of them contain extensive bodies of pyritized quartzite that contain small and variable amounts of copper, silver, and gold.

Like the ore deposits in dolomite rocks, the deposits in the Rocky Point zone of the Sawatch Quartzite are related in occurrence to a combination of solution channels, discussed under "Rock Alteration," and fractures, discussed under "Structure." Unlike the deposits in

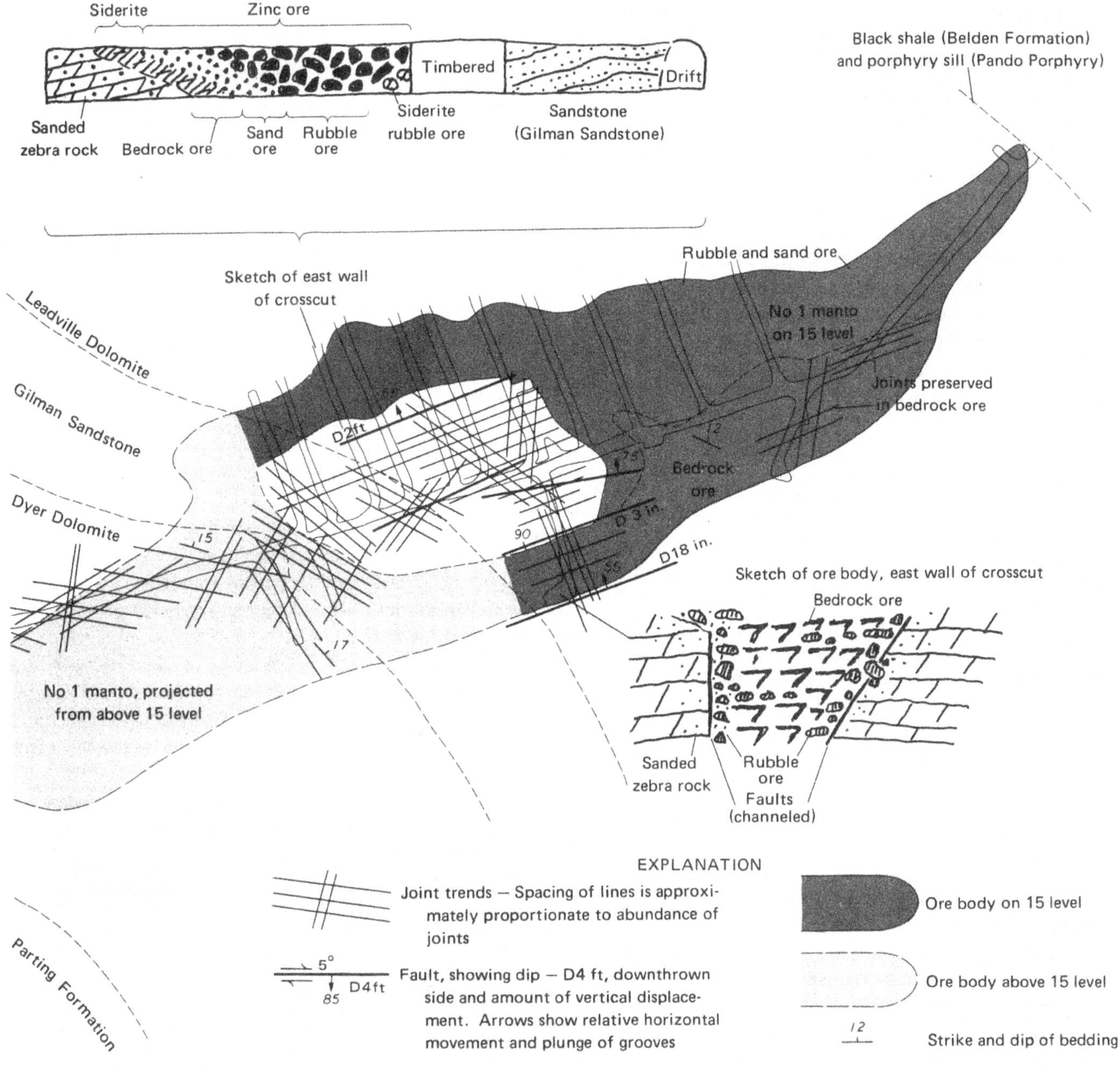

dolomite rocks, however, those in the quartzite are a combination of replacement and fracture- or cavity-filling deposits.

Six general types of ores or mineralized materials constitute most of the deposits in the Rocky Point zone, and the ore bodies of the Bleak House mine — seemingly a special case — constitute a seventh type. From oldest to youngest and, in the oxidized and enriched zones, from deeper to shallower, these are (1) an early and widespread pyrite (pyrite I) that is nearly barren except where secondarily enriched; (2) a siderite–pyrite II–marcasite–chalcopyrite–sphalerite assemblage (hereafter referred to as sideritic ore) that contains values in gold and silver and is younger than the pyrite I mineralization; (3) auriferous pyrite III, some of it very rich, with which silver-gold tellurides (not seen by us) may have been associated; (4) secondarily enriched sulfide ores, valuable for copper, silver, and gold; (5) iron oxides that contain gold and silver; and (6) a puttylike clay (hereafter referred to as argillic ore) that contains horn silver (cerargyrite) and gold. In addition, the Bleak House mine, the only mine in the quartzite that has produced lead and zinc, contains a body of argentiferous galena-sphalerite ore.

The deposits containing these various ores are distributed in northeast-trending belts, each of which is the site of a major mine or small group of mines (fig. 11). Between the belts, mineralization was scant. Except where oxidized, the belts are broad pyritized zones within which are more localized and sharply defined subbelts of sideritic deposits and still more localized deposits of auriferous pyrite III. Each of the belts coincides with a system of anastomosing solution channels that extends up the dip of the quartzite in the Rocky Point zone. The channel systems are most strongly developed in the southeast part of the district, particularly in the Ground Hog mine. Many of the mine workings follow channels that either constituted small pipelike ore bodies or sites of easy excavation.

The various types of deposits are discussed under subsequent headings, and then some of the principal mines are described.

EARLY PYRITE DEPOSITS

The widespread early pyrite in the mineralized belts now marked by the principal mines in the Rocky Point zone is classed as pyrite I. Deposition of this pyrite clearly preceded that of the sideritic ore in quartzite, as discussed under the next heading. The pyrite I mineralization stage was essentially monomineralic. No other hypogene ore mineral is evident in most bodies of the pyrite, though in a few places a very little chalcopyrite or black sphalerite may be seen.

In the pyritized belts, which are as much as 1,000 ft (≈300 m) wide, pyrite is distributed in certain beds of the quartzite within the Rocky Point zone and also occurs widely as a fracture filling. In the mineralized beds, pyrite replaced quartz in both fresh and altered quartzites; it also replaced the sericite and clays in altered quartzites, and, in the lenses of dolomitic quartzite present in the northern part of the district, it replaced the dolomite cement. The degree of replacement varies from bed to bed and area to area. In some beds or areas, disseminated pyrite constitutes only a few percent of the rock; in others, pyrite constitutes more than half the rock, or even replaces the rock totally. The number and the stratigraphic position of the pyritized beds differ from area to area. In some places, only a single bed about a foot (30 cm) thick is pyritized; in others, several pyritized beds are distributed through stratigraphic sections as much as 20 feet (6 m) thick. In most localities, one or two beds are persistently pyritized over wide areas, whereas other beds are pyritized only near fractures (fig. 46) or solution channels.

Pyrite deposited in open spaces occurs abundantly as veinlets or veins along joints and small faults, as seams or flat veins along bedding faults, and as a filling in quartzite breccia (fig. 47). It occurs also as a filling in small solution channels or as an encrustation on the walls of larger channels.

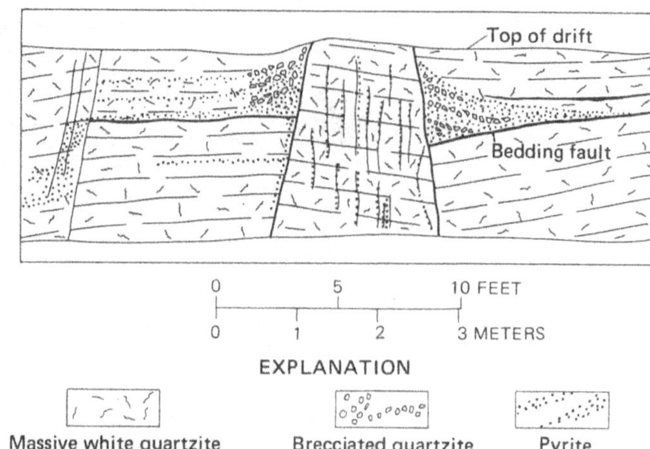

FIGURE 46.—Relation of pyrite to steep faults and flat faults in quartzite, Rocky Point mine. Typical tongue-fault block in center.

Though no other ore minerals are visible in most occurrences, the early pyrite contains small amounts of gold, silver, and copper and thus constitutes a resource of very low grade. Assay data from various sources indicate a gold content from nil or a trace to as much as $1 per ton (at the $20.67 price of gold); a silver content from a trace to about 2 ounces per ton; and a copper content from nil to as much as 0.4 percent. All the higher values reported are from vein (or veinlet) occur-

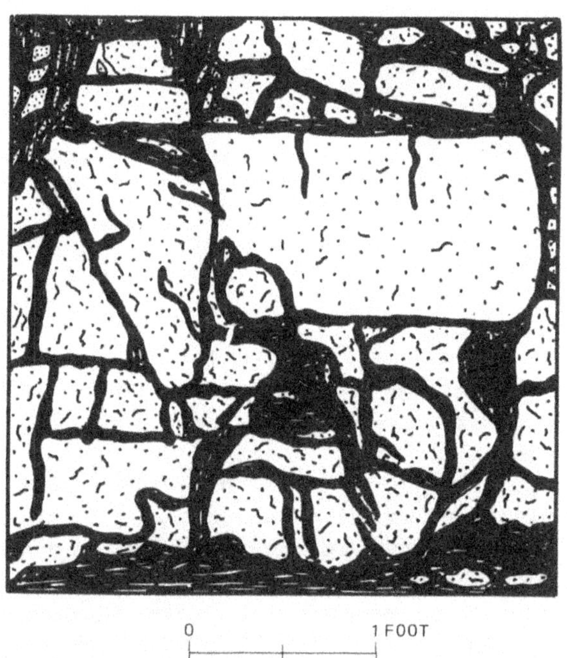

FIGURE 47.—Pyrite veinlets (dark) in brecciated quartzite, Rocky Point mine.

rences; the bedded replacement deposits seem to be barren or nearly so except near veins. The contents of base and precious metals in the early pyrite rise near the contacts with sideritic ore, and they are, of course, also affected by secondary enrichment. Thus, as it exists now, the early pyrite is of varied composition.

The association of pyritized rock with solution channels indicates that the two features are closely related. Channels clearly served as conduits for pyrite-depositing solutions, as shown by shells of pyritized rock surrounding some channels and by a fill or lining of pyrite in others. On the other hand, some channels cut indiscriminately back and forth through pyritized and fresh quartzite and evidently postdate pyrite deposition. From these relations, we conclude that dissolution of quartzite began before the deposition of pyrite, but continued during and after deposition, and that dissolution of quartzite and deposition of pyrite I were joint effects of a single surge of hydrothermal solutions.

SIDERITIC ORE

Sideritic ore consists of the minerals siderite, pyrite II, marcasite, black sphalerite (marmatite), chalcopyrite, galena, and barite, and of associated gold and silver in an unidentified form. The proportions of these minerals vary widely. Some deposits, as in the Champion mine (fig. 11; pl. 2*B*), consist almost entirely of siderite. Many others consist of siderite, subordinate pyrite, and small and variable amounts of the other sulfide minerals. Marcasite is a major constituent of one ore body in the Percy Chester mine but is only a minor accessory mineral in others, as in the Rocky Point mine. Chalcopyrite is widespread but sparse. In a few places in the Rocky Point mine, it is accompanied by very sparse tetrahedrite. Black sphalerite is likewise widespread but sparse; it is nowhere abundant enough to be mined except in the Bleak House mine. Galena occurs only in scattered localities and is very sparse except in the Bleak House mine. Gold and silver contents of the ores are erratic. Ores that contain chalcopyrite generally contain gold and silver, but some that consist megascopically only of siderite and pyrite (or marcasite) also contain gold and silver. On the whole, the sideritic ores are low in grade (except in the Bleak House mine), and they have been mined principally in the zone of secondary sulfide enrichment.

The sideritic ores occur only in restricted and sharply defined areas or belts within the broad pyritized belts discussed under the preceding heading. When the distribution of siderite and the minerals that accompany it is plotted (pl. 2*B*–*D*), the sideritic belts emerge as channel-shaped areas that split and flare in the updip direction like river distributaries on a delta. These areas consist only in part of bodies of siderite ore; in large parts of them, siderite and associated minerals are only sparsely distributed as thin coatings or scattered crystals on surfaces of quartzite or pyritized quartzite or as veinlets in a few fractures of certain orientations in these rocks. The zoning in barite crystals scattered sparsely on surfaces of siderite ore — discussed under "Minerals of the Ore Deposits" — indicates that the ore-depositing solutions moved updip along a tortuous route in the siderite-bearing belts or channelways. These belts are easily traced through the oxidized zone by means of the barite crystals — which survived oxidation — and also by means of pseudomorphs of siderite crystals preserved in iron and manganese oxides.

Sideritic ore occurs chiefly as a filling or coating in solution channels in either fresh or pyritized quartzite (fig. 48); small amounts are in veinlets along steep fractures within the Rocky Point zone, and some is in replacement bodies in the walls of solution channels (fig. 49). Thus, most sideritic ore bodies have the shapes of pipes of various sizes (several inches to several feet in diameter). The pipes are very irregular, and they constrict or enlarge abruptly. They are generally flattened in the plane of the bedding, and in places where dissolution spread laterally along bedding-fault breccias, they widen into mantolike bodies. A replacement manto unique to the Bleak House mine is described under the "Bleak House" heading.

Sideritic ore coats pyritized rock, fills veins or veinlets that cut pyritized rock, and locally replaced pyritized rock. Thus, it defines a mineralization stage that not only followed the stage of pyrite I and the dissolution of quartzite but was of different character. Except for differences in proportions, sideritic ore has the mineralogy of the manto deposits in dolomite rocks, and we correlate it with those deposits both in time and in the character of the mineralizing solutions.

AURIFEROUS PYRITE III

Pyrite that is rich in gold occurs in small pockets and veinlets scattered through some of the nearly barren early pyrite deposits. We refer the auriferous pyrite to pyrite III, though as noted under "Minerals of the Ore Deposits," correlation with the copper-silver ores of the chimney ore bodies in dolomite rocks cannot be proved. Auriferous pyrite is especially characteristic of the Ground Hog mine (pl. 2*A*), where it is the chief material mined below the oxidized zone. In this and other mines, small pockets, pipes, and veins of the auriferous pyrite were mined selectively through openings barely large enough for a man to crawl in, or even to admit his arm. Consequently, very little of the pyrite remains to view.

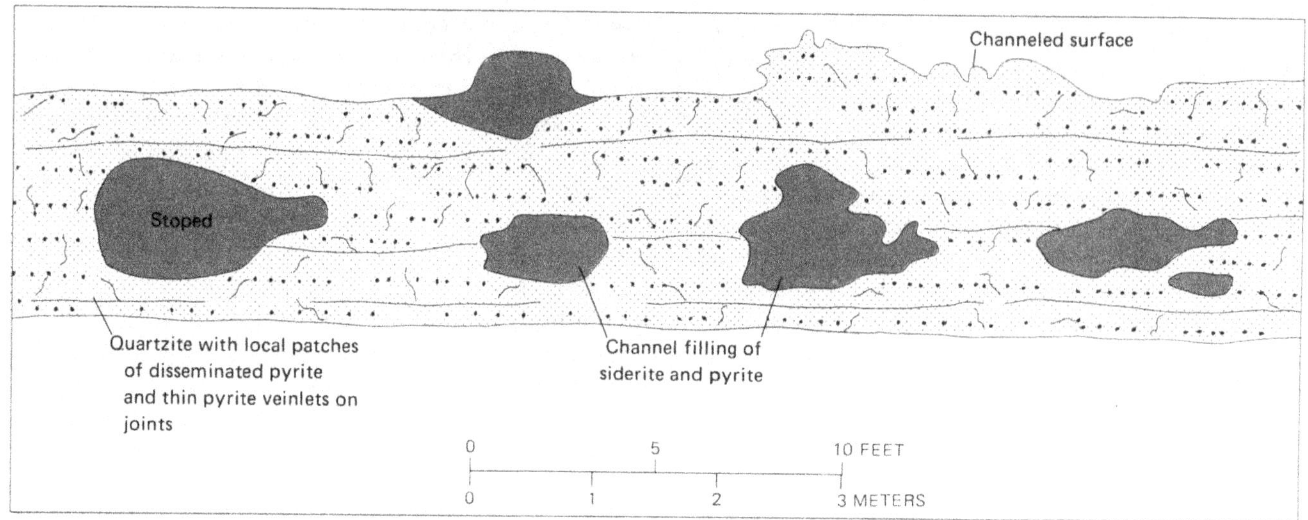

FIGURE 48. — Wall of drift in Rocky Point mine, showing small solution channels in quartzite filled by sideritic ore. Each pocket of ore has an opening that extends downdip (into the page); on opposite side of drift, the ore-filled channels merge into a channel ore body 4 ft (1.2 m) high and several feet wide.

As judged from the few scraps remaining, the auriferous pyrite is distinguished from the nearly barren early pyrite (pyrite I) by a generally coarser grain and by octahedral and cubic crystal habits, as contrasted to pyritohedral. Similar features were noted by Means (1915, p. 16) to be in the auriferous pyrite vein worked through the Doddridge winze in the Ground Hog mine (pl. 2*A*).

Though telluride minerals were not seen by us in the ore deposits in the Rocky Point zone, we infer from the locations of telluride occurrences reported in the past that these minerals are associated with the auriferous pyrite. Telluride minerals may have been the principal repository of the silver in scattered pockets of pyrite ore mined in the past that were rich in silver as well as in gold.

ENRICHED SULFIDE ORES

Except in the Bleak House and Ground Hog mines, most of the mining below the oxidized zone has been confined to the zone of secondary sulfide enrichment. In this zone, copper and silver — and possibly some gold — that had been leached from the oxidized zone were redeposited in the preexisting sulfide ores. By this process, early pyrite and sideritic deposits of inherently low grade were enriched sufficiently to make them mineable. In the enriched ores, chalcocite and minor covellite and bornite occur as coatings and minute veinlets on and in pyrite. Argentite (or acanthite) is mixed in powdery form with some of the chalcocite and locally occurs also in discrete grains between pyrite grains. A single grain of pyrargyrite was observed in one of several samples of enriched ore studied microscopically by us.

The enriched ores are typically in streaks within bodies of early pyrite or sideritic ore, and they constitute only small ore bodies — in many cases, little wider than a mine drift.

OXIDIZED ORES

By far the largest part of the gold and silver produced from the ore deposits in the Sawatch Quartzite came from the oxidized zone. Two types of gold-silver ore are present in this zone: (1) iron oxide, or mixtures of iron and manganese oxides and iron sulfates, mainly in the lower part of the oxidized zone, and (2) argillic ore, in the upper part of the oxidized zone. No sharp boundary exists between the areas occupied by the two ore types. The two occur together in places, and long fingers of the iron oxide type project updip to or near the surface along the trends of some of the original sideritic deposits.

Abundant iron oxide material in blanketlike bodies as much as 6 ft (2 m) thick remains in the old mine workings in the oxidized zone. This material evidently was too low in grade to be mined. As judged by the workings within these bodies, the material workable for its gold and silver content is in pipelike or ribbonlike bodies extending down the dip of the quartzite. The pipelike bodies are at the base of the iron oxide blanket and are irregular in size and shape. Except where they flare on the bedding in thin zones at their base, most of them are less than 6 ft (2 m) in diameter, and some are as little as 1 ft (30 cm) in diameter. The ribbonlike bodies are only a few inches or a foot (30 cm) thick and lie either at the base or at the top of the iron oxide blanket. Some are as much as 50 ft (15 m) wide in

FIGURE 49. — Sideritic ore (dark area) that has replaced Sawatch Quartzite, Rocky Point mine. Light rim of dark area is secondary efflorescence of iron sulfates. Area of photograph is about 2 × 4 ft (0.6 × 1.2 m).

places. They were worked in "belly stopes" as little as 18 in. (45 cm) high.

As described by Guiterman (1891), gold in some of the iron oxide ore occurs as nuggets or as aggregates of grains cemented by horn silver and iron sulfates. In other ore, the gold is not visible and is detectable only by assay. Guiterman reported that lumps of iron sulfates that assayed high in gold revealed no gold "colors" upon crushing and panning. In our observation, the iron oxide material that was mined and is presumably gold bearing generally contains iron sulfates, though not all iron-sulfate-bearing material is auriferous. Stratified ocher (fig. 39), which evidently formed by sedimentation of transported iron oxide and iron sulfate particles, was not mined anywhere. Clinkery aggregates of iron and manganese oxides that characterize the upper parts of the iron oxide blankets are almost devoid of iron sulfate minerals and were not mined.

Argillic ore is an aggregate of clays and aluminous material — mainly alunite — that contains gold in discrete grains and small nuggets and in aggregates cemented by horn silver. The ore occurs as blanket seams only a few inches thick but of wide extent. Where iron oxides are present, such seams generally overlie the iron oxides. Elsewhere, the seams have a floor that is a bedding plane above hard, little-altered quartzite. In both occurrences, the quartzite above the ore seam is generally severely leached and spongy. The sponginess is due in part to the removal of disseminated pyrite and in part to the dissolution of silica. The gold and silver in the argillic seam are concentrated near the base, especially in depressions in the slightly wavy floor (Guiterman, 1891). The metals also occur in cracks in the quartzite floor. Discarded mining paraphernalia in "belly stopes" indicate that mining was accomplished with implements as small as table knives, nut picks, and whisk brooms and that the ore was dragged for long distances to haulageways in sacks or powder (dynamite) boxes. The powder boxes bear dates in the 1880's and, in a few places, the early 1920's.

Argillic ore is the end product of leaching that was so intense that even the iron and manganese of the original pyrite and siderite were removed. Additionally, one or more beds of quartzite locally were largely or entirely dissolved away. The gold and silver in the residual argillic ore were concentrated from several sources: the small amounts in early pyrite, the somewhat greater but still small amounts in sideritic ore, the gold in scattered veinlets and pockets of auriferous pyrite III, and, probably, scattered veinlets and small pockets of silver-gold telluride minerals. We see no reason to posit as sources for the rich oxidized ores parent deposits any richer in gold and silver than those now preserved downdip from the oxidized zone.

BLEAK HOUSE MINE

The Bleak House mine is the northernmost major mine in the Sawatch Quartzite (fig. 11). Ore deposits of the mine are in the Bleak House vein and in a replacement manto that straddles the vein in the Rocky Point zone (fig. 50). The vein, which contains argentiferous galena, was worked on a small scale in the early years of mining in the district, prior to 1900. The updip part of the manto ore body, which consists of argentiferous lead-zinc ore, was worked together with the vein in the

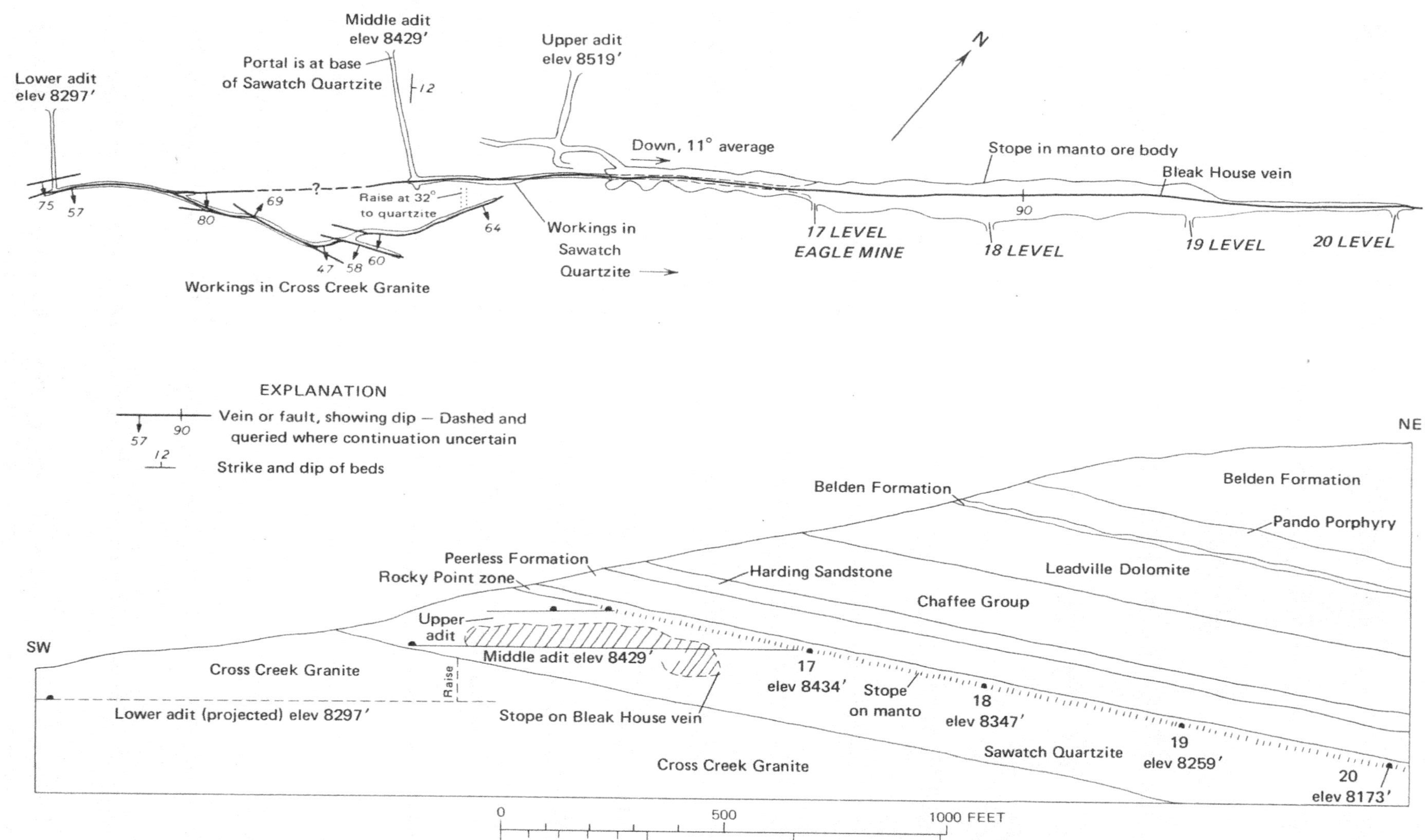

FIGURE 50. —Sketch map and longitudinal section of the Bleak House mine. Map is constructed from several sources and is in part diagrammatic.

approximate period 1910–15. The lower and major part of the manto was mined in the early 1960's through workings extended from the Eagle mine. At the time of our studies in the 1940's the old workings were inaccessible, except for the nonproductive lower adit in Precambrian rocks (fig. 50). Our information on the mine is based on the report made by Means (1915, p. 13–15), on unpublished maps made by J. M. Dismant, operator of the mine in 1910–15 period, on an unpublished map made by A. J. Hiester, and on a brief examination made by us in 1964 of the empty stope in the lower part of the manto ore body.

The workings of the mine consist of three adits and the long inclined stope on the manto ore body (fig. 50). Portals of the adits are near the bottom of the canyon of Rock Creek (pl. 1). All three adits consist of crosscuts southeastward to the Bleak House vein or related fractures and of drifts on the northeast-trending vein or fractures. The lower adit is not connected with the other workings, and according to a note on the map by A. J. Hiester, its location with respect to the other workings is not precisely established. The middle adit follows the Bleak House vein northeastward to the intersection with the inclined stope in the Rocky Point zone. The vein was stoped extensively above the adit level and in smaller degree below (fig. 50). The upper adit opens into the stope on the combined Bleak House vein and manto ore body.

The Bleak House vein is well defined and mineralized only in the Sawatch Quartzite. It apparently dies out upward in the Peerless Formation, and its counterpart in the granite below the Sawatch Quartzite is a zone of interesecting discontinuous faults and fractures (fig. 50). In the quartzite, the vein strikes about N. 50° E. and is vertical. It occupies a fault along which the northwest side is displaced downward 2–3 ft (61–91 cm). As described by Means (1915, p. 13), the vein in the old workings is as much as 2 ft (61 cm) wide and consists of argentiferous galena, sphalerite, minor pyrite, and quartzite fragments. As seen by us in the lower and newer workings, it consists of stringers of galena in a sheeted zone about 2 ft (61 cm) wide. Means (1915, p. 25) noted that the vein is indistinguishable in parts of the manto ore body. We conclude from this that the fault occupied by the Bleak House vein existed prior to mineralization in the manto, a view contrary to that of Means. According to Crawford and Gibson (1925, p. 61), Dismant reported that galena ore in parts of the vein is sheared and slickensided, indicating some post-ore fault movement. This is the only example of such movement known to us in the entire Gilman district.

The manto ore body occupies a stratigraphic zone 6–8 ft (1.8–2.4 m) thick that consists of interbedded quartzite and dolomitic sandstone lying a few feet below the top of the Sawatch Quartzite. The dolomitic sandstone is partly or largely replaced by black sphalerite, argentiferous galena, pyrite, and siderite. The quartzite beds, which are somewhat sericitized, are less extensively replaced. Because of the galena and its contained silver, the ore of the manto in quartzite has greater value than that of the manto ore bodies in dolomite rocks. The manto is 75–100 ft (23–30 m) wide in its lower part and tapers updip to a width of 50–60 ft (15–18 m). As seen on 18 and 19 levels of the Eagle mine, it is bordered on at least the southeast side by open solution channels beyond which the rock is only moderately pyritized. The manto bottoms abruptly just below 19 level, though a narrow zone along the Bleak House vein was worked down to 20 level. Large open solution channels extend downdip from the bottom of the ore body. The walls of these channels are locally coated with thin mammillary crusts of very fine grained light-yellow sphalerite.

Means (1915, p. 13) stated that the Bleak House vein is continuous from the Sawatch Quartzite into the Precambrian granite and that the dip changes abruptly at the contact from vertical in the quartzite to 55° SE. (misstated as "NE") in the granite. This statement is evidently based on the single exposure in the raise from the lower adit to the base of the quartzite (fig. 50). Our map, however, indicates faults or fractures of various trends and dips in the granite below the quartzite. One fault that might be the Bleak House dips 80° SE., not 55° (fig. 50). Displacements on the faults in the granite are consistently left lateral, and grooves and slickensides consistently plunge northeast at low to moderate angles. Thus, a fracture zone at least 200 ft (61 m) wide and characterized by subhorizontal fault movements in the granite gives way upward to a single small fault (the Bleak House vein) in the quartzite. The plane of separation is at the base of the quartzite, which is the site of a persistent bedding fault, as discussed under "Structure."

The faults in the Precambrian rocks of the lower adit have altered walls and are locally mineralized in minor degree. Veinlets of black sphalerite and pyrite, with or without quartz, are fairly common. Chalcopyrite and bornite are present locally. Galena is rare. Marcasite was observed with sphalerite and pyrite at the junction of the entrance crosscut and the drift and was also observed on the dump. Wallrock alteration is similar to that in the Whipsaw mine, discussed under "Rock alteration." The Whipsaw (pl. 1) is on the trend of the Bleak House fracture zone, just across the Eagle River, to the southwest. This vein strikes N. 40°–50° E., dips steeply southeast, and shows subhorizontal (strike-slip) fault movement. It consists of 1–3 ft (30–90 cm) of sheared, sericitized, and locally silicified granite that contains stringers of pyrite and black sphalerite. A caved stope indicates some production, probably of silver and gold.

ROCKY POINT MINE

The Rocky Point mine is in the Sawatch Quartzite on the south side of the promontory on which the town of Gilman is located (pl. 1), above the Newhouse tunnel and beneath the updip part of the No. 1 manto in the Leadville Dolomite (fig. 11). The mine (pl. 2*D*) consists of a maze of workings in the plane of the bedding in the Rocky Point zone, near the top of the quartzite, as in the Bleak House mine (fig. 50, section). The mine has several portals in the cliffs south of Gilman; these were reached in the past by ladders and cable tramways but at the time of our studies they were no longer readily accessible. At its lower or northeast end, the main incline reaches the level of the Newhouse tunnel and is connected to that tunnel by a short drift (pl. 2*D*).

The Rocky Point was one of the first mines operated in the Gilman district, and most of its gold and silver production predates 1900. The lower part of the mine — in the enriched sulfide zone — was worked again on a small scale in the 1930's. Though no record of production is available, the mine probably was the most productive of the gold-silver mines in the quartzite, as judged from the extent of the stopes. Production possibly was in the range of \$1–\$2 million.

The mineral deposits of the mine are localized in or adjacent to quartzite beds that are shattered or brecciated as a result of bedding-fault movements. The bedding breccias are typically about 2 ft (61 cm) thick, though they swell locally to as much as 5 ft (1.5 m). In most places, only one breccia bed is present, but in some, as many as three beds are distributed through a stratigraphic thickness of 10–12 ft (3–3.6 m). In the bedding-fault movements that created the bedding breccias, the upper beds moved generally north (N. 10° W. to N. 20° E.). Rocks above and below the bedding breccias are strongly jointed and are cut in many places by steep northeast- and northwest-trending fractures, some of which displace the bedding a fraction of an inch. Though individually nonpersistent, the northeast-trending fractures define a belt of fracturing that coincides in general with the main ore trend (pl. 2*D*).

Solution channels (fig. 22) are abundant throughout the mine. They occur principally in the stratigraphic zone occupied by bedding breccias and mineralized quartzite. They follow the bedding and also steep fractures (pl. 2*D*). The presence of dolomite sand in a channel along a fracture at the bottom of the mine indicates that at least some of the channels connected somehow with channels in the Dyer or Leadville Dolomite. Much of the sideritic ore in the mine occurs as a filling in channels (fig. 48).

The ores produced from the mine were all oxidized or secondarily enriched and were mainly derived from sideritic ore, which is distributed in channellike belts that split and flare in the updip direction (pl. 2*D*). The asymmetry of zoned barite crystals indicates a flow of solutions in the updip direction, as does also the pattern of the sideritic deposits.

The upper half of the mine is in the oxidized zone and consists principally of extensive stopes of very low ceiling. Most of the workings shown on the map of this part of the mine are merely rock-cribbed passageways through stopes. The stopes are on both argillic and iron oxide ores, principally on the argillic ore in the uppermost workings and on the iron oxide ore at greater depth. The uppermost iron oxide ores are clinkery and porous and are coated with silt that contains micas washed down from the surface. Depressions in the silt-coated iron oxide are filled with travertine, which is horizontally banded, indicating deposition from pools of standing water.

In the lower half of the mine, the width of the area mineralized with siderite is narrower, and only small areas are stoped. The stopes are in partly oxidized and secondarly enriched sideritic deposits and early pyrite in replacement beds. Such ores contain copper as well as silver and gold.

Assays of grab samples of various types of ores are listed in table 8; locations are shown on plate 2*D*. The samples listed as chalcopyrite, sphalerite, and galena ores each contain conspicuous amounts of the mineral named but are representative of only small pockets of ore. They are intended chiefly to show the occurrence of gold and silver in sulfide ore that was the progenitor of the rich oxidized ores.

PERCY CHESTER MINE

The portal of the Percy Chester[1] mine is high on the quartzite cliffs about 1,800 ft (550 m) southeast of the portal of the Newhouse tunnel (fig. 11). According to a former owner, the mine was opened in 1883, and by 1923 it had produced about \$1 million in gold, silver, and copper (Crawford and Gibson, 1925, p. 81). The mine has not been worked since the 1920's.

The workings (pl. 2*C*) consist of a crooked incline about 1,100 ft (335 m) long and many lateral inclined drifts and unmapped low stopes in the plane of the Rocky Point zone of the Sawatch Quartzite, as in the Bleak House mine (fig. 50, section). The workings connect to the southeast with workings off the Wheeler incline of the Pine Martin mine. This incline reaches the

[1] Spelled Pursey Chester on some maps and by Crawford and Gibson (1925, p. 81).

TABLE 8. — *Assays of ore samples from Rocky Point mine*

[Assays by New Jersey Zinc Co., Gilman, Colo. Locations of samples shown on pl. 2D]

Sample No.	Character and occurrence	Ounces per ton		Percent		
		Au	Ag	Cu	Pb	Zn
1	Pyrite veinlets in quartzite, in lowest workings. Possibly pyrite III?	0.10	8.7	0.2	0.25	0.2
2	Chalcopyrite and pyrite in sideritic ore, bottom of main incline.	.16	8.1	3.8	.65	.75
3	Chalcopyrite ore, occurrence as No. 2.	.34	60.0	11.0	1.0	.8
4	Black sphalerite ore, occurrence as No. 2.	.06	6.1	.8	6.05	29.1
5	Enriched granular pyrite I, lowermost northwest stope.	.03	2.1	.2	.2	.55
6	Chalcopyrite ore in sideritic deposit, location as No. 5.	1.20	160.0	15.7	1.0	.2
7	Galena ore, location as No. 5	.16	28.6	.3	37.9	.3
8	Sulfate efflorescence on pyritic quartzite.	.05	4.1	.1	1.0	.35
9	Gray secondary sulfide minerals in argillic material in oxide-sulfide transition zone.	.10	10.8	Nil	42.4	.8
10	Argillic seam below No. 9	.02	1.8	.1	1.05	.4
11	White argillic ore characteristic of oxidized zone.	.32	26.5	Nil	.45	.5
12	Barite-bearing argillic ore in oxide-sulfide transition zone.	.08	8.2	2.6	.5	.25

surface in the face of a sheer cliff more than 100 ft (30 m) high and nowadays is inaccessible except through the Percy Chester mine. The Percy Chester and Wheeler inclines were served at one time by very steep cable tramways to the railroad in the canyon bottom, but this means of ore transport was later supplanted by a route through the Mabel mine. By this route, ore was dropped through two successive winzes, one of 28 ft (8.5 m) and one of 191 ft (58.2 m), to the third level of the Mabel (pl. 2*C*); thence, it was trammed to the Mabel shaft, hoisted 145 ft (44 m), and then lowered to the railroad via a short cable tramway. Obviously, the ore had to be of good grade to finance such extensive handling.

Most of the ore deposits in the Percy Chester mine are in thin-bedded gray quartzite just below the top of the Sawatch Quartzite, stratigraphically a few feet higher than in the Rocky Point mine. The quartzite is brecciated in one to three bedding zones within a 4- to 6-ft (1.2- to 1.8-m) thickness of the quartzite. In many places, only vague outlines of breccia fragments remain after pyritization and local sericitization of the quartzite. A lower mineralized zone, 28 ft (8.5 m) below the main one, was worked on the Star Chamber sublevel (pl. 2*C*). This zone was not seen by us but is inferred to be in crackle breccia in thick-bedded vitreous white quartzite.

The ore bodies in the mine are in and along solution channels that trend generally parallel to the dip of the quartzite. The main channels are in the brecciated bedding zone and are localized along the intersections of steep northeast-trending fractures with this zone (pl. 2*C*). Subsidiary channels and ore deposits occur along northwest-trending fractures.

A large part of the output from the mine was in secondarily enriched sulfide ore closely associated with siderite. Stopes in the sulfide zone are the most extensive among the gold-silver mines in the Sawatch Quartzite. The sideritic deposits in the northwestern part of the mine area are in narrow solution channels along fractures, and the deposits in the southeastern part are in a broader, sinuous channel for which little fracture control was seen during a brief examination (pl. 2*C*). In both areas, the sideritic bodies cut through pyritized quartzite and beds that were almost completely replaced by early pyrite (pyrite I). In the northwestern area, the siderite is accompanied by pyrite II, and in the southeastern area, by marcasite or by this mineral plus pyrite II. Additionally, pyrite III may be present in both areas. The relations among the iron sulfides in this mine are complex and warrant further study if opportunity should allow. However, when the mine was mapped in 1946, the poor quality of the air in the lower workings discouraged detailed study.

In the northwestern area, along the Percy Chester incline, extensive stopes were made in a bed of massive pyrite I 3–5 ft (0.9–1.5 m) thick where this bed is cut by fractures and solution channels containing sideritic ore. The stopes extend for several feet on one or both sides of the narrow sideritic bodies (pl. 2*C*), and the pyrite remaining in the stope walls is streaked with a black coating of secondary chalcocite and bornite(?). Sulfide ore that presumably came in part at least from these stopes is reported to have averaged 0.41 ounce gold and 12 ounces silver per ton and 4 to 5 percent copper (Crawford and Gibson, 1925, p. 81). However, according to an old assay map of unknown origin, the pyrite remaining in the walls of the stopes contains only about a tenth as much gold, silver, and copper as the reported ore grade. Evidently, gold, silver, and copper were added to the bed of early, low-grade pyrite in the vicinity of the sideritic veins and channel fills. In many places, no mineralogic expression of this addition is evident. In some, however, pyrite of different character than the early pyrite is present. For example, early fine-grained pyrite I of pyritohedral habit is coated or veined in places by coarse pyrite II and III(?) (crystals as large as 1 in. (2.5 cm)) of other crystal habits. The habits of the coarse crystals differ from stope to stope: pyritohedrons in one, combinations of cubes and dodecahedrons in another, and still other combinations of crystal forms in others. In a few places the bed of early pyrite is pitted by solution cavities or small channels, and some of these are lined with the coarse crystals or are filled with white clay (kaolinite-dickite) that contains pyrite crystals.

In addition to pyrite, the sideritic ore along and near the Percy Chester incline contains small amounts of black sphalerite and chalcopyrite and a very little galena. The siderite itself is of two types. The predominating type, and the only one present in most places, is the brownish-gray variety in cockscomb crystals that is typical of the entire Gilman district. The other type is dull pink and in rhombic crystals. The pink color suggests a high manganese content, but where oxidized, the rhombic crystals alter only to iron oxide. Where the two siderites occur together, the cockscomb variety rests on the rhombic variety. A grab sample of the rhombic variety assayed 0.36 ounce gold and 11.64 ounces silver per ton and 0.08 percent copper. These metals presumably are contained in minute sulfide inclusions.

The marcasite-bearing sideritic ore in the southeastern part of the mine occurs principally as a filling or coating in solution channels of various sizes clustered in a sinuous belt as outlined on plate 2*C*. The channels are cut in pyritized quartzite and beds or lenses of pyrite I. A few of the channels are partly filled with a younger coarsely crystalline pyrite, and this pyrite is in turn overlain by the siderite-marcasite assemblage. In the channels containing this assemblage, siderite and marcasite are reciprocally distributed. In the deepest workings, marcasite is abundant and siderite is sparse. In going updip, the amount of marcasite decreases as that of siderite increases. Evidence of marcasite — as recorded by pseudomorphs of its distinctive crystals (fig. 30) — disappears in the lower part of the oxidized zone. The siderite and marcasite are accompanied by small amounts of black sphalerite and crystals of quartz. Marcasite is locally coated with minute crystals of chalcopyrite and, in at least one sample, with arsenopyrite (fig. 31). It is also coated by clots of white kaolinite (fig. 30). A grab sample of marcasite assayed 1.24 ounces gold and 6.0 ounces silver per ton and 0.02 percent copper. If representative, this sample would indicate fairly rich ore by the standards of the 1970's. It should be noted, however, that the marcasite occurs only as crusts on the walls of solution channels, many of which are only a few inches in diameter.

In the oxidized zone broad low stopes were made in argillic ore, and long fingerlike stopes were made in iron oxide ore. A sample from a box of argillic ore left in one of the stopes near the base of the oxidized zone assayed 0.41 ounce gold and 5.08 ounces silver per ton. The iron oxide ore that was stoped is on the trends of the sideritic or marcasitic deposits deeper in the mine. Marcasite-bearing ore is oxidized to greater depth in the mine than is pyrite. Thus, in a belt about 200 ft (61 m) wide at the base of the oxidized zone, fingerlike bodies of iron oxide along former marcasite-siderite channels project into the pyritic deposits. The oxidized ores are reported to have averaged about $150 per ton in gold and silver, and pockets were found with values as high as $2,000 per ton (Crawford and Gibson, 1925, p. 81).

GROUND HOG MINE

The Ground Hog mine, in the southern part of the productive area (fig. 11), consists of several inclines and extensive lateral workings in an area extending about 1,300 ft (396 m) along strike of the Sawatch Quartzite and as much as 1,500 ft (457 m) down the dip (pl. 2*A*). The mine was opened in 1884 (Tilden, 1887) and was operated more or less continuously until the early 1920's. Production records are lacking for the period prior to 1898. Crawford and Gibson (1925, p. 71) reported a production of $660,000 in gold and silver as compiled from incomplete records. The total may have been twice that.

The ore deposits of the Ground Hog mine are in the Rocky Point zone of the Sawatch Quartzite, as in the Bleak House mine (fig. 50, section). The chief mineralized zone is in slightly to moderately brecciated white quartzite 10–18 ft (3–5.5 m) below the base of the Peerless Formation. Locally, thin-bedded gray quartzite above the white quartzite is mineralized, and in a few places basal strata of the Peerless Formation are mineralized for a few feet outward from fractures. The most mineralized parts of the "ore beds" are along steep northeast-trending fractures, which are abundant in the mine (pl. 2*A*). Many of the fractures are limited vertically to a few feet of quartzite beds between bedding faults; grooves on the walls of the fractures plunge about parallel to the dip of the quartzite, and, thus, the fractures are minor tongue faults. A few fractures have greater vertical extent. The Cleveland fault, which is followed by the Doddridge incline (pl. 2*A*), extends upward into the Peerless Formation and downward beneath the Rocky Point zone an unknown distance. This fault has a normal-fault dipslip displacement of about 3 ft (1 m), with the southeast side down. A sheeted zone containing the vein worked in the Doddridge winze was explored downward all the way to the granite; no displacement is evident along the sheeted zone as exposed at the head of the winze.

Solution channels are abundant in the mine. They range in size from tubes only a few inches in diameter (fig. 23) to smoothly domed cavities as much as 20 ft (6 m) in diameter in the lower part of the Badger incline. Concentric spall structure (fig. 23) is pronounced along some channels and weak or absent along others. Most of the channels follow the northeast-trending fractures, but some meander through beds of crackle breccia in an average northeast-southwest direction.

The ore deposits occur as pipelike fillings or linings in solution channels, as narrow veins along the fractures, and as narrow manto replacement deposits in certain quartzite beds along the fractures. All the ore bodies are small in cross section, generally little wider than a drift or incline and, in many places, much narrower. The mine lacks the broad low stopes that characterize all the other mines in the Sawatch Quartzite. The ore produced came from the drifts and inclines and, in places, from stopes on veins. Most of the veins are less than 4 in. (10 cm) wide, and most of the stopes on them extend no more than 10 ft (3 m) below and 20 ft (6 m) above the inclines. In the Doddridge winze, a vein was worked on at least two levels, at depths of 35 and 85 ft (10.6 and 25.9 m) below the incline.

The primary ore in the Ground Hog mine consisted almost entirely of pyrite before oxidation and enrichment. Siderite and the base-metal sulfides that accompany it are very sparse. The pyrite exposed in the workings in the sulfide zone is varied in character and in content of precious metals. Pyrites I, II, and III are all present in the mine, but in many places where distinctive crystals are lacking, as in veins, the identity of the pyrite is uncertain. Fine-grained pyritohedral pyrite I in pyritized quartzite was not mined except in narrow belts along veins or fractures, where it was evidently enriched at the pyrite II or pyrite III stages (as well as, possibly, by secondary enrichment). Where pyrite was mined, coarser and younger pyritohedral pyrite is present, typically as crystals 0.25–0.5 in. (6–13 mm) in diameter. In some places, the coarse pyrite is accompanied by small amounts of black sphalerite and galena and by traces of siderite. Therefore, the coarse pyritohedral variety is classed as pyrite II. Octahedral and cubic pyrite, in crystals as large as 1 in. (2.5 cm) in diameter, is classed as pyrite III. This pyrite was mined wherever exposed, and only traces of it remain to view. Assays of grab samples of a few types of pyrite are listed in table 9.

TABLE 9. — *Assays of ore samples from Ground Hog mine*

[Assays were made by a commercial firm. Locations of samples shown on pl. 2*A*; n.d., not determined]

Sample No.	Character and occurrence	Ounces per ton	
		Au	Ag
1	Pyrite II(?) crystals in white clay in cavity in pyritized quartzite.	0.08	3.04
2	Coarse pyrite II filling channels cut in fine-grained pyrite I.	.56	15.32
3	Pyritized basal bed of Peerless Formation, in raise. Pyrite I.	.02	1.16
4	Coarse pyrite II intergrown with galena.	1.18	n.d.
5	Granular pyrite I(?) in bed adjoining a vein that contains small amounts of siderite.	.38	3.28

The most productive areas in the mine are along the Doddridge and Nottingham inclines and the vicinity of the Doddridge winze (pl. 2*A*). The Badger incline is in an intervening area of nearly barren quartzite, except at its lower end. The Doddridge incline is in a strongly fractured and mineralized belt along the Cleveland fault. In the oxidized zone, the incline follows an irregular pipe of iron oxide, and most of the numerous lateral workings follow smaller iron oxide pipes, some less than 2 ft (61 cm) in diameter. In the sulfide zone, the incline and lateral workings follow pyrite veins and channel fillings of coarse pyrite. Ore deposits along the other inclines are generally similar to those of the Doddridge. A pocket of rich telluride ore described as hessite was found near the intersection of the Forgy and Nottingham inclines (Engineer and Mining Journal, 1894, v. 51, p. 540). Though this is the only occurrence of telluride minerals reported from the mine, it seems likely that undetected telluride minerals and associated free gold account for the richness of some small bodies of pyrite ore. Means (1915, p. 17) reported that no free gold was detected at high magnification in polished sections of pyrite crystals in ore from the Doddridge winze that contained 21.10 ounces gold and 39.15 ounces silver per ton. He does not indicate whether inclusions other than gold were present in the crystals, or whether any coatings or intergranular materials — such as telluride minerals — were present.

The grades of the ores mined in the Ground Hog probably varied widely. Sulfide ore mined in the lower part of the Doddridge incline in the early 1920's was worth $25–$125 per ton (at the $20.67 price of gold) according to a former lessee. Guiterman (1891) stated that oxidized ore from the Ground Hog contained 7 ounces gold and 50 ounces silver per ton, but this must apply to selected ore. The gold and silver content of iron oxide ranges down almost to zero, and much material poorer in grade than that reported by Guiterman probably was mined.

CHAMPION MINE

The Champion mine, the southernmost of the major mines in the Sawatch Quartzite (fig. 11), consists of a main incline that extends 2,000 ft (610 m) down the dip of the quartzite and a maze of lateral workings along the upper 1,500 ft (457 m) of the incline (pl. *2B*). The mine was opened in 1880 or 1881 and was operated discontinuously until about 1920. No record of production is available, but extensive stopes suggest an output of gold and silver generally comparable to that of the Percy Chester and Ground Hog mines.

The ore deposits and workings in the upper part of the mine are in gray quartzite of the Sawatch 5–10 ft (1.5–3 m) below the base of the Peerless Formation. In the lower part of the mine, the main workings are in white quartzite of the Sawatch 30–40 ft (9–12 m) below the base of the Peerless. There, the upper ore zone was explored or worked through numerous raises. At both stratigraphic levels, quartzite beds in zones 1–4 ft (0.3–1.2 m) thick are brecciated and pyritized in varying degrees. Steep fractures and small faults, mostly of northeast trend (pl. *2B*), are much less numerous than in the Ground Hog mine (pl. *2A*). Some of the fractures terminate upward or downward at the bedding faults marked by the brecciated beds, and some extend through the brecciated beds. Many of the fractures are followed by solution channels within the brecciated and pyritized bedding zones.

The ore deposits are characterized by abundant siderite, in marked contrast with the deposits in the adjoining Ground Hog mine. The siderite occurs as small mantos as much as 4 ft (1.2 m) thick, as a filling in solution channels, and as veinlets in fractures. The mantos are replacement deposits in the walls of solution channels, and they generally have knife-edge outer contacts along fractures. The individual mantos and channel-fill deposits are only a few feet wide. They are clustered in a sharply defined belt that widens and splays out updip as shown on plate *2B*. Part of a second belt is exposed in the northwest part of the mine. The siderite of this belt is principally the pink rhombic variety that occurs also in the Percy Chester mine. An analysis (table 4) indicates a very low manganese content.

The siderite bodies contain fine-grained pyrite and locally abundant barite crystals, but no visible sphalerite, galena, or chalcopyrite. In general, only the channel-filling siderite was mined; it is inferred that the pyrite in such siderite is accompanied by silver and gold. The pyrite in the replacement siderite of the mantos evidently is largely pyrite I inherited from the preexisting pyritized quartzite. Assay data that we have seen indicates that the replacement siderite contains only traces of gold and an ounce or two of silver per ton. In the area where pink rhombic siderite occurs, pyrite in some small areas is coated by secondary chalcocite. This indicates a minor copper content in the sideritic material, though no primary copper minerals were observed.

In the oxidized zone, both argillic and iron oxide ores were mined. The argillic ore was mined in broad low stopes, and the iron oxide (actually, iron-manganese oxide) in flat pipes along the trends of original channel-fill siderite deposits. Large amounts of iron-manganese oxide derived from siderite of replacement origin still remain.

TELLURIDE DEPOSITS

Very small silver-gold telluride deposits occur in the lower part of the Sawatch Quartzite in the vicinity of the Ben Butler, Star of the West, and Tip Top mines (pl. *2E*). The deposits were discovered and worked in the course of mining along sulfide veins in the Precambrian rocks, though the veins terminate upward at the base of the quartzite (fig. 5). The telluride deposits occur as films and thin veinlets on joint cracks and bedding planes in the quartzite in scattered small areas above the sulfide veins. In places the veinlets widen into small pockets that yield a few ounces to as much as 100 lbs. (45 kg) of telluride minerals. The deposits have been mined as high as 35 ft (11 m) above the base of the quartzite in very crooked inclined raises barely large enough to admit a man. The telluride ore consists of hessite veined by petzite which in turn is veined by free gold (fig. 36). No sulfide or gangue minerals are present, and the quartzite host rock is unaltered.

VEINS IN PRECAMBRIAN ROCKS

GENERAL FEATURES

Fissure veins in Precambrian rocks have been worked or explored in several mines along the canyon bottom below the ore deposits in Sawatch Quartzite and Leadville Dolomite. Many of the mines produced only small amounts of ore, which was valuable principally for gold. Most of the mines were inaccessible in the 1940's, and none has been worked since then. The only ones studied in detail were the Ben Butler mine and adjoining workings (pl. 2*E*). The Mabel mine, probably the most productive of the mines in the Precambrian rocks, was described briefly by Crawford and Gibson (1925, p. 67–70).

The veins in Precambrian rocks all strike northeast and dip steeply. Many of them occupy fracture zones a few inches to a few feet wide within broad Precambrian shear zones that are elements of the Homestake shear zone, as discussed under "Structure." The fracture zones end abruptly at the bedding fault at the base of the Sawatch Quartzite (fig. 5), or they pinch markedly at the fault (fig. 6) and die out in the lower part of the quartzite. Fault movement along the fracture zones was subparallel to the gentle dip of the overlying quartzite. The mineralized streaks in most of the veins are a few inches to 2 ft (61 cm) wide and consist principally of quartz and pyrite. Black sphalerite, galena, chalcopyrite, and traces of tetrahedrite are present in small ore shoots along the quartz-pyrite veins. Ore consisting of these minerals and pyrite contains gold and silver. Ore produced from the Mabel mine contained 2–4 ounces gold and 5–15 ounces silver per ton as well as 8–10 percent lead and as much as 3 percent copper (Crawford and Gibson, 1925, p. 67). Wallrocks of the veins are strongly altered. (Discussed in section entitled "Rock Alteration.") Disseminated pyrite in the sericitized and silicified rocks adjoining the veins is referred to pyrite I. Some pyrite in parts of the veins that consist only of quartz and pyrite (and were not mined) may also be pyrite I. The pyrite associated with the gold- and silver-bearing base-metal sulfides is interpreted to be pyrite II, though no siderite — which normally is associated with pyrite II — is present in the veins. We have no data on the gold and silver content of the pyrite II but infer that the precious metals are principally in chalcopyrite and galena, as in the Rocky Point mine (table 8). Pyrrhotite, which is locally present in small amounts, is paragenetically younger than the pyrite II and older than the black sphalerite in the sulfide ore. In a few places, coarse crystals of pyrite rest on intergrowths of black sphalerite and pyrite II and seem to have been selectively mined. The coarse and presumably auriferous crystals may be pyrite III. They are coated locally by crystals of pale-amber sphalerite, the youngest mineral in the veins.

Unlike the ore deposits in the overlying sedimentary rocks, those in the veins are unoxidized, though some just below the base of the quartzite show signs of secondary enrichment. The lack of oxidation reflects two factors: (1) the youth of the canyon, which was cut after the first and, in part, during the second and third of several glacial episodes (Tweto and Lovering, 1977), and (2) the absence in the Precambrian rocks of the highly permeable solution channel systems so widespread in the mineralized parts of the Sawatch Quartzite and the Leadville Dolomite.

BEN BUTLER AND ADJOINING MINES

The Ben Butler mine, at the Belden railroad siding (fig. 11), consists of two sets of workings, one in the Precambrian rocks and basal Sawatch Quartzite and one in the Rocky Point zone near the top of the quartzite. The upper set was largely inaccessible in 1941 and was not mapped. The workings in Precambrian rocks (pl. 2*E*) connect with workings on the Star of the West and Tip Top claims to the south. The main Star of the West mine is in the Rocky Point zone of the Sawatch Quartzite, but a small unit of it follows the underside of the quartzite. Lateral workings of the Tip Top adit are on some of the veins of the Ben Butler mine. The Tip Top adit crosscuts 1,200 ft (366 m) through Precambrian rocks and ends in basal beds of the Sawatch Quartzite.

The Ben Butler lower adit follows a tight, virtually barren fault that strikes about N. 70° E. and is nearly vertical, in migmatitic granite. This fault has a left-lateral displacement of 45 ft (14 m); grooves on its walls plunge 10° NE. The fault was explored upward in a raise to the base of the quartzite, where it ends. However, telluride deposits occur on joints in the quartzite above the fault and were mined in irregular wormholelike workings.

Near its inner end, the adit intersects three northeast-trending veins (pl. 2*E*). All three veins are in sheared granite and migmatite of a Precambrian shear zone. The sheared rocks contain abundant fine-grained green biotite and, as a consequence, are dark greenish gray. In the walls of the veins, this rock is bleached, sericitized or silicified, and locally pyritized. The veins consist of 6 in. to 4 ft (15 to 122 cm) of fractured, silicified rock cut by veinlets of pyrite and quartz and by seams of gouge. All have been stoped irregularly, presumably for their gold content. Where the northwestern vein reaches the Sawatch Quartzite, it rolls over and ends in a thick gouge seam in the bedding fault at the base of the quartzite (fig. 5). A small body of pyritic ore was mined immediately under the gouge. A few thin veinlets of pyrite cut the quartzite above the gouge; these veinlets extend only a few feet vertically and then bend into bedding planes and die out.

Two of the veins in the Ben Butler were worked from the Tip Top adit (pl. 2*E*). Small stopes, most of them less than 30 ft (9 m) high, or deep, are spaced intermittently along the veins. Between the stopes, the veins are nearly barren of sulfide minerals. At 940 ft (286.5 m) from the portal of the Tip Top adit, a northeast-trending vein was stoped upward into the quartzite. This vein, which consists of stringers of gold- and silver-bearing pyrite, sphalerite, and galena, dies out upward about 10 ft (3 m) above the base of the quartzite. Above the vein, in an area that is about 20 ft (6 m) in diameter and that extends upward to a stratigraphic level 30 ft (9 m) above the base of the quartzite, telluride minerals in thin veinlets on bedding planes and joints were mined. As exposed near the inner end of the Tip Top adit, the base of the Sawatch Quartzite is a smooth surface that has a 1- to 4-in. (2.5- to 10-cm) seam of sulfide ore (pyrite, black sphalerite, galena, and trace of chalcopyrite) pasted to it. This seam is underlain by 6–24 in. (15–61 cm) of gougy bedding-fault breccia that consists of altered migmatite, beneath which is silicified rock that contains disseminated pyrite.

The lower workings of the Star of the West mine (pl. 2*E*) consist mainly of a short incline that follows the intersection of a northeast-trending vein in the granite with the base of the quartzite. This vein is one of the few that faults the base of the quartzite, displacing it 4 or 5 ft (1.2 or 1.5 m) (fig. 6). In the granite, the vein is 5–10 ft (1.5–3 m) wide and consists of gougy breccia cut by veinlets of sulfides. The breccia and its walls consist of sheared rock in a Precambrian shear zone. In the quartzite, the vein is only a single tight fracture or narrow sheeted zone. The granite and the quartzite are separated by 1–5 ft (0.3–1.5 m) of gouge.

Two kinds of ore — sulfide and telluride — were mined in these small workings. Sulfide ore was obtained mainly from a pyritic seam on the underside of the quartzite; in places bedding seams were also mined in the lower few feet of the quartzite, and seams in the granite on fractures subparallel to the base of the quartzite were also mined. In a short sublevel at the foot of the incline, the vein itself was stoped. There, it contains streaks 1–6 in. (2.5–15 cm) wide of coarse pyrite crystals, resinous sphalerite, and quartz crystals. Like the coarse pyrite in some of the mines in the quartzite, this pyrite evidently contains gold. It is referred to pyrite III. The telluride ore was obtained from thin veinlets and scattered small pockets on joints in the quartzite in small areas above the workings on the vein. The "wormhole" workings driven for telluride ore follow one joint or bedding plane and then another to a maximum height of about 35 ft (11 m) above the base of the quartzite.

SUMMARY OF HYDROTHERMAL HISTORY

The age relations between the various kinds of altered rocks and ore deposits in the three different hostrock settings in the Gilman district are summarized in table 10. The geologic age of the hydrothermal events, the conditions under which the hydrothermal processes operated, and the character of the processes are discussed in following sections.

By our interpretation, hydrothermal activity began with dolomitization of the Leadville Limestone. Later, parts of the dolomite rock formed by this process recrystallized to forms such as zebra rock and pearly dolomite. Both the dolomitization and the recrystallization occurred throughout an extensive part of the mineral belt and thus they are not specifically related in origin to the Gilman center of mineralization, though they did provide a receptive host rock for mineralization. All varieties of recrystallized dolomite were in existence before the stage of dissolution and mineralization began. The onset of dissolution marks a profound change in the hydrothermal regime. This change probably was abrupt, and we infer that the recrystallization of dolomite continued, at least on a minor scale, up to the moment of change.

The mineral deposits and related rock-alteration features of the district fall into two age groups, which we relate to early and late mineralization stages (table 10). In the early stage, dissolution and rock alteration were intense, and base-metal ores containing small but variable amounts of precious metals were deposited widely. In the late stage, ores valuable principally for precious metals were deposited locally, mainly within the older ore bodies. Little or no rock alteration accompanied this stage of mineralization except for the local replacement of dolomite by jasperoid at Red Cliff.

The early mineralization stage began with dissolution in the Leadville and Dyer Dolomites and the intervening thin, dolomitic Gilman Sandstone, with dissolution in the Rocky Point zone of the Sawatch Quartzite, and with wallrock alteration along veins in the Precambrian rocks. Deposition of sulfides began after many dissolution features had been formed but before dissolution had ceased. Thus, dissolution and sulfide deposition were closely related in time, and they were

TABLE 10. — *Succession and correlation of hydrothermal events in three different hostrock environments*

	Dolomite environment	Quartzite environment	Precambrian-rock environment
Late mineralization stage	Remineralization of pyrite I–II cores of chimney ore bodies with Cu, Ag, Au, Pb, Sb, Bi, Te minerals; deposition of Ag-Pb ore and small bodies of jasperoid at Red Cliff.	Deposition of auriferous pyrite III and silver-gold telluride minerals in small pockets and veinlets.	Deposition of auriferous pyrite III and resinous sphalerite as veinlets and coatings.
	Interruption in mineralization, accompanied or followed by some dissolution of pyrite in chimneys.	Interruption in mineralization.	Interruption in mineralization.
Early mineralization stage	Deposition of base-metal ores (siderite–pyrite II–sphalerite assemblage) in mantos and chimneys; argillic alteration in karst silt and porphyry.	Deposition of sideritic ore (siderite, pyrite II, marcasite, and base-metal sulfides) as filling or lining in solution channels and local replacement deposits.	Deposition of base-metal ores in small ore shoots in veins; late argillic (dickite) alteration in wallrocks.
	Continued dissolution. Interruption in mineralization. Deposition of pyrite I in "protochimneys"; sericitic alteration of porphyry.	Continued dissolution. Interruption in mineralization. Deposition of pyrite I as replacement of quartzite and fracture filling; local mild sericitic alteration.	Continued alteration; probable interruption in mineralization. Deposition of pyrite I contemporaneously with sericitization and silicification of wall rocks.
	Dissolution (sanding, channeling, and deposition of cave rubble and dolomite sand).	Dissolution (channeling) and local mild argillic alteration.	Argillic alteration in wallrocks of veins.
Dolomite stage	Recrystallization of dolomite. Dolomitization of limestone.	Not affected.	Not affected.

evidently accomplished jointly by solutions that changed progressively both with time and with distance of travel in the host rocks.

In the dolomite rocks, the hydrothermal solutions that caused dissolution and subsequent mineralization circulated initially in a combination of ancient (pre-Belden) karst solution channels and the fractures and bedding faults of later (Laramide) origin. The karst channels were in the upper Leadville and probably in the Gilman Sandstone. The hydrothermal solutions enlarged, extended, and integrated the karst openings in the Leadville, and they severely altered the Gilman. They produced sanded (friable and porous) rock along channels and fractures, complex systems of channels and caves, and fillings of resedimented dolomite sand and cave rubble in the openings. In the Sawatch Quartzite, the solutions were guided by bedding faults and steep fractures (joints and tongue faults). The solutions dissolved quartzite completely, leaving open channels devoid of rubble or sand. Quartzite in certain beds in the walls of the channels and of some fractures was first mildly argillized and then mildly sericitized. In the Precambrian rocks, where dissolution features are absent, the early solutions produced zones of argillic alteration in the walls of the veins.

Deposition of sulfide minerals began while dissolution was in progress and continued after it ended. The first step in the sulfide stage was the deposition of pyrite I in all three hostrock environments — dolomite rock, quartzite, and Precambrian rocks. This pyrite was accompanied only by traces of other sulfide minerals and precious metals. In the Sawatch Quartzite, where the relation of pyrite I to other assemblages is clearest, the pyrite replaced quartzite in brecciated beds and in the walls of solution channels and fractures; it was deposited also as a filling in fractures and solution channels. In the dolomite rocks, evidence of the pyrite I stage of mineralization is masked by the voluminous products of later stages. We infer from rather sketchy data that pyrite I was deposited as "protochimneys" at the sites of the present pyritic cores of the chimneys and at the lower ends of the manto ore bodies. In the veins in Precambrian rocks, pyrite I was deposited in disseminated form in the sericitized and silicified rocks adjoining the vein walls and, with quartz, as a filling in narrow and essentially barren veins.

The second step in the early mineralization stage was the deposition of the base-metal ores. These ores are characterized principally by siderite, pyrite II, and black sphalerite (marmatite); but, depending on setting, they also contain economically significant galena, chalcopyrite, and precious metals, and genetically significant marcasite and pyrrhotite. In the dolomite rocks, the base-metal ores replaced dolomite to form the manto ore bodies and the mineralogically zoned enlargements of the pyrite I bodies in the chimneys.

Dissolution of dolomite rock accompanied mineral deposition, though on a dwindling scale with passing time. Illitic clay in karst silt in pockets in the dolomite rock was replaced by kaolinite and dickite. Sericitized porphyry in the sill above the Leadville also underwent alteration to kaolinite and dickite. In the Sawatch Quartzite, the base-metal ores were deposited on a much smaller scale than in the dolomite rocks. Except in the Bleak House mine, the base-metal ores were deposited principally as a filling or lining in solution channels; some, however, replaced pyritized quartzite of the pyrite I stage. In the Bleak House, a moderately large manto was formed by replacement of dolomitic sandstone beds bordering the Bleak House vein. In general, the base-metal ores in quartzite contained more chalcopyrite, gold, and silver than those in the dolomite rocks. In the veins in Precambrian rocks, the base-metal ores were deposited in small ore shoots consisting of pyrite, chalcopyrite, galena, and black sphalerite in various proportions and containing gold and silver. Late dickite formed in the altered wallrocks adjoining the veins at the same time.

In some places, marcasite was deposited at an early stage in the base-metal mineralization. As judged from the deposits in Sawatch Quartzite, the marcasite was deposited contemporaneously with siderite and before the base-metal sulfides and before most — if not all — pyrite II. As inferred from ore texture, marcasite was also deposited in the lower parts of the manto ore bodies in dolomite rocks and was later converted to pyrite II. Small amounts of pyrrhotite were deposited with pyrite II and black sphalerite in the chimney ore bodies and contiguous parts of the mantos in dolomite rocks, and also in the veins in Precambrian rocks.

The change in mineralization from pyrite I alone to the siderite–marcasite–pyrite II–base-metal assemblage was abrupt. No evidence of a transition is seen, and, in the quartzite, much dissolution occurred after deposition of pyrite I and before deposition of the sideritic ores. Hydrothermal activity in the form of dissolution probably was continuous from the pyrite I stage to the sideritic stage, but an interruption in mineralization seems evident. We have no data to appraise the duration of this interruption but doubt that it was lengthy.

In the late mineralization stage, ores rich in precious metals were deposited in volumes much smaller than those of the base-metal ores in all three hostrock environments in the Gilman area and in the Leadville Dolomite at Red Cliff. In the chimney ore bodies at Gilman, parts of the pyrite I–pyrite II cores of the chimneys were remineralized with silver- and gold-bearing copper and lead sulfide and sulfosalt minerals and silver-gold telluride minerals, forming copper-silver ore. Siderite, pyrite, and sphalerite — the three main constituents of the base-metal ores — were deposited only sparsely. The pyrite is classed as pyrite III; the sphalerite is a resinous iron-poor variety that differs markedly from the iron-rich black sphalerite of the early mineralization stage. At Red Cliff, argentiferous galena, resinous sphalerite, and pyrite, presumed to be pyrite III, were deposited in scattered small ore bodies. Gold and a bismuth mineral were deposited locally. Dolomite rock was replaced by jasperoid in small bodies in the vicinity of the ore deposits, and in places the rock was veined by jasperoid that contains trace amounts of silver and copper.

In the Sawatch Quartzite only auriferous pyrite III and silver-gold telluride minerals were deposited in the late mineralization stage. Pyrite III was deposited in small amounts as coarse crystals in openings or small veins in quartzite, pyrite I, or sideritic ore. Telluride minerals were deposited in very small quantity in local pockets associated with pyrite III in the Rocky Point zone and as veinlets unaccompanied by any other mineral in the lower part of the Sawatch. In the veins in Precambrian rocks, pyrite III and resinous sphalerite were deposited locally in veinlets and as a coating on pyrite II–black sphalerite ore.

As seen best in the chimney ore bodies, the minerals of the late mineralization stage were deposited principally as vug fillings and encrustations in porous older pyrite (a composite of pyrites I and II). The porosity of the older pyrite was caused by dissolutions prior to deposition of the late minerals.

ONE MINERALIZATION EPISODE OR TWO?

The marked contrast between the ores of the early and late mineralization stages and the absence of any transitional facies indicate an interruption in the mineralizing process. The interruption can be interpreted as (1) a brief break in deposition associated with a sudden change in the character of the mineralizing solutions, or (2) a lengthy break between two separate episodes of mineralization. A brief break in a single major episode of mineralization has been assumed implicitly in the past (Lovering and Tweto, 1944; Radabaugh and others, 1968; Titley, 1968). However, the recognition in recent years of successive episodes of mineralization in the region surrounding Gilman (discussed under "Age of Mineralization") suggests the possibility that the early and late mineralization stages represent separate episodes of mineralization and are of appreciably different geologic ages. This possibility has a bearing on deductions as to depth of cover and pressure conditions during mineralization, discussed in later sections.

Differences between the early and late mineralization stages may be summarized as follows: The early mineralization stage was characterized by abundant iron, sulfur, and zinc, small amounts of lead, and very small amounts of copper, silver, and gold. The late mineralization stage, in contrast, was characterized by copper, silver, gold, and lead in combination with antimony, bismuth, and tellurium, as well as sulfur. Zinc was present only in a small amount in this later stage, and iron was sparse, as indicated by the sparsity of pyrite III and siderite. The abundant sphalerite of the early stage was a black iron-rich variety, whereas the sparse sphalerite of the late stage was a resinous iron-poor variety. Minerals of the early stage were deposited principally by replacement, whereas those of the late stage were deposited principally as encrustations, vug fillings, and veinlets cutting the older ores. The mineralogy of the early stage suggests deposition at moderately high temperatures, as discussed in a following section. In all, the ores of the early stage are mesothermal in aspect, and those of the late stage are epithermal in aspect.

AGE OF MINERALIZATION

No direct measure of the age of the ore deposits in the Gilman district is available. The deposits have long been correlated with Laramide igneous activity in the mineral belt, and, hence, classed as Laramide in age (Crawford, 1924; Burbank and Lovering, 1933; Radabaugh and others, 1968). Though a Laramide age is a reasonable inference for some or all of the deposits at Gilman as well as in the Leadville, Kokomo, Breckenridge, Alma, and Aspen districts, proof of Laramide age is lacking. What is known in all these districts is that the ore deposits are younger than most faults and folds and younger than the Laramide igneous rocks exposed in the districts. No minerals of the ore deposits have been dated, and unfortunately, minerals that might yield reliable dates — such as adularia — have not been found in the deposits. Sericite in altered rocks associated with the ore deposits might yield meaningful dates but presents problems both because of argon loss in fine-grained sericite and because of the frequent occurrence of sericite of more than one generation. Sericite in altered Pando Porphyry at Leadville (Linn, 1963, p. 44–45) and at Pando (T. G. Lovering, unpub. data) has yielded K-Ar ages of 60 m.y. Unfortunately, the Pando Porphyry was sericitized in two stages, one deuteric, immediately after crystallization 70 m.y. ago, and one later, presumably at the mineralization stage. If any of the older sericite remained in the samples that were analyzed — as it probably did — the 60 m.y. date would be spuriously high. If argon loss occurred, the date might be spuriously low, depending on the proportions of the two sericites and whether there were differential losses. An attempt to use the rubidium-strontium method to date sericite from the Pando at Leadville and the molybdenum ore body at Climax yielded ages of 100±11 and 73±8 m.y., respectively (Moorbath and others, 1967, p. 232). These ages are utterly inconsistent with the geologic facts.

Recognition in recent years of igneous and mineralization episodes distinctly younger than Laramide (<40 m.y. in age) compounds the problem of fixing the age of ore deposits that are known only to be younger than certain Laramide igneous and structural features. At Climax, molybdenum-tungsten mineralization occurred in three pulses correlated with three pulses of intrusion in a granitic stock about 30 m.y. old (Oligocene), and this mineralization was followed by a sparse base-metal--rhodochrosite--fluorite mineralization (Wallace and others, 1968). In the Gore Range, minor or trace mineralization occurred in the absence of visible igneous features in Miocene or later time and, in the form of fluorspar, in as late as Pliocene or Pleistocene time (Tweto and others, 1970, p. 75). In the southern Sawatch Range, mineralization occurred after intrusion of the Mount Princeton batholith, and on the western side of the Front Range, after intrusion of the Montezuma stock, both dated as Oligocene in age (C. E. Hedge, oral commun., 1973). In more distant areas, mineralization occurred after emplacement of porphyry stocks 10–12 m.y. old (Mutschler, 1970; Obradovich and others, 1969; Segerstrom and Young, 1972).

In the absence of definitive data, we conclude that the ore deposits of the early base-metal mineralization stage at Gilman are of Laramide age and, more specifically, probably near 60 m.y. (Paleocene) in age. The deposits of the late mineralization stage might be of approximately this same age or, as noted by one of us (Tweto, 1968b), they might be appreciably younger. If younger, they are most likely of Oligocene age because Oligocene intrusive rocks and mineral deposits are much more abundant and widespread in the mineral belt than those of still younger Tertiary age. Except for the sill of Pando Porphyry, no igneous rocks are exposed near Gilman. The nearest igneous bodies are several miles to the southeast, in the Pando, Kokomo, and Climax areas, where Laramide and Oligocene intrusives occur together (Tweto, 1974; Tweto and Case, 1972, p. 4–7; Wallace and others, 1968).

DEPTH OF MINERALIZATION

The depth at which the ore deposits formed is measured geologically by the thickness of rocks inferred to have been present above the Leadville Dolomite at the time of mineralization. At present, the cover over the Leadville ranges from zero at the canyon of the Eagle River to 5,500 ft (1,677 m) 6 mi (9.7 km) northeast of Gilman (Tweto and Lovering, 1977, pl. 1, sec. *E–E′*). As judged from regional stratigraphic and paleotectonic relations, the Leadville in the Gilman area was covered by about 18,000 ft ($\approx$5,500 m) of younger sedimentary rocks at the beginning of Laramide orogeny in late Late Cretaceous time, when the area was at or near sea level. By the time of mineralization in Paleocene time, some part of the original cover had been removed as a result of uplift and erosion in the Sawatch and Gore Ranges, but the loss probably had been partly compensated by sedimentation in the trough between the two ranges. The problem is to evaluate the losses and gains.

The great anticline of the Sawatch Range began to rise in late Campanian Cretaceous time about 72 m.y. ago and continued to grow both vertically and laterally through the end of the Cretaceous (65 m.y. ago) and into the Paleocene (Tweto, 1975). The Gore Range began to rise slightly later, in the period 70–65 m.y. ago, and continued to grow in the Paleocene. Both ranges were eroded as they rose, and concurrently they shed sediments to bordering sedimentary basins. Erosion was greatest in the older and structurally highest parts of the ranges and was only nominal in the younger and peripheral parts. The Gilman area is on the edge of the trough between the northern tip of the northward-plunging Sawatch Range anticline and the southern tip of the Gore Range (fig. 3). Thus, it is in the area of the youngest parts of these uplifts. In such a setting, Laramide erosion probably was not severe even in the uplifts, and in the trough between the uplifts it could have been largely compensated by sedimentation.

As the White River Plateau (fig. 1) did not begin to rise until Eocene time (Tweto, 1975), the inference may logically be made that a huge sedimentary basin that occupied most of northwest Colorado in Laramide time extended into the trough between the northern Sawatch and southern Gore Ranges. If so, sedimentation — rather than erosion — probably occurred in the trough during the early stages of uplift in the mountain ranges. The sediments so produced would have been equivalents of parts of the uppermost Cretaceous Lance Formation and the Paleocene Fort Union Formation. The amount of such sediments may not have been large in this embayment on the periphery of the main sedimentary basin — possibly no more than 1,000 ft (300 m). However, the accumulation and then the removal of the Laramide sediments could have occupied much of the time between the beginning of Laramide orogeny and the mineralization stage.

We conclude that by the time of mineralization about 60 m.y. ago, the 18,000-foot cover above the Leadville — mentioned previously — was largely intact in the area just east of Gilman, though reduced somewhat. We adopt a thickness value of 15,000 ft (4,575 m) of cover over the bottom of the trough, about 5 mi (8 km) northeast of Gilman, and 12,000 ft (3,660 m) at Gilman. From Gilman southwestward toward the crest of the Sawatch anticline, the thickness presumably decreased rapidly. Thus, a pressure gradient decreasing from northeast to southwest existed in the Gilman area at the early mineralization stage.

The depth of the Leadville in Oligocene time — if the late mineralization stage occurred then — is less definite. Erosion probably proceeded slowly in the trough between the ranges through Eocene time, and it may even have been interrupted by deposition of sediments equivalent to a part of the Eocene Wasatch Formation. The fact that the Dakota Sandstone, the basal unit of Cretaceous rocks, is still preserved in the little-deformed trough area in the Minturn quadrangle 13 mi (21 km) north of Gilman (Tweto and Lovering, 1977, pl. 1) together with structural relations suggests that by the end of Eocene time, 38 m.y. ago, erosion in the Gilman area may have extended little deeper than the base of the Mesozoic rocks. Thus, the cover at Gilman may have been reduced to 8,000–10,000 ft (2,440–3,050 m). Little net change is inferred during the part of Oligocene time that preceded Oligocene mineralization. The area probably was repeatedly veneered with volcanic tuffs and sediments produced during the great volcanic episode in Colorado in Oligocene time (Steven, 1975). Altitude and relief were low at the time in most of the mountain area of the State, and, at most, erosion probably only kept pace with deposition.

ISOTOPIC COMPOSITIONS OF SULFUR AND LEAD

Isotopic data on sulfur in sphalerite and associated pyrite in two samples of manto ore as reported by Ault and Kulp (1960, p. 86) are shown in table 11. Though the data are limited, they indicate only small variations in S^{34} content and a grouping that — compared with the sulfur in some ore deposits — is close to zero per mil. Such features are characteristic of ore deposits associated with intrusive igneous rocks in many western mining districts (Jensen, 1967, p. 148, 152–153), though Rye and Ohmoto (1974) have shown that exceptions occur elsewhere. The sulfur isotope

TABLE 11. — *Isotopic analyses of sulfur in pyrite and sphalerite from Gilman*

[Data from Ault and Kulp (1960, p. 86)]

	Percent Fe in sphalerite	S^{32}/S^{34}	S^{34} per mil
Sample 1:			
Sphalerite	9.9	22.14	+3.2
Pyrite		22.14	+3.2
Sample 2:			
Sphalerite	9.6	22.16	+2.3
Sphalerite	1.3	22.18	+1.4
Pyrite		22.17	+1.8

values in the Gilman ores are closely similar to those in the ores of Leadville (S^{34}, -0.9 to +0.9 in galena (Ault and Kulp, 1960, p. 86)), which are clearly in igneous association (Tweto, 1968a, p. 703). We interpret the isotopic data to indicate a magmatic source for the sulfur in the Gilman ore deposits. We once considered the possibility that the sulfur might have been derived somehow from the huge reservoir of sulfur in the Pennsylvanian evaporite deposits nearby. However, sulfur in evaporite deposits is generally heavier and more variable isotopically than the sulfur in the ore deposits (Ault and Kulp, 1960, p. 75). The possibility that the ore sulfur was derived from the Pennsylvanian evaporites would seem to be precluded.

Isotopic data on lead in ore and in carbonate rocks in the Gilman area are shown in table 12. A marked difference between the ore leads (Nos. 1–7) and the rock leads (Nos. 8–10) indicates that the ore leads were not mobilized from their carbonate host rocks. Among the ore leads, no significant differences in isotopic compositions of leads in manto, chimney, and vein ores are evident. Leads of these same general compositions occur in scattered veins in the Sawatch Range southwest of Gilman and also in the Leadville and Kokomo districts and occur in ore deposits in middle Tertiary intrusive rocks at Climax and in the Sawatch and Front Ranges (Delevaux and others, 1966, table 1). According to Bruce Doe (oral commun., 1974), leads of these same compositions characterize many of the major mining districts of the West. Thus, the lead at Gilman — like the sulfur — fits a pattern characteristic of many western mining districts associated with igneous rocks. It differs markedly from the lead of the Mississippi Valley-type mineral deposits (Heyl and others, 1974).

HYDROTHERMAL PROCESSES

We conclude this report with a discussion of hydrothermal processes as inferred from the character and setting of the ore deposits. The conclusions are necessarily empirical, as we lack the isotopic, fluid-inclusion, and detailed mineralogic data that nowadays provide sharp definition of many aspects of hydrothermal processes. We recognize three major hydrothermal stages (table 10). The dolomite stage — the first — reflects the influence of magmatic heat in moderate amount but little or no addition of magmatic materials. The early mineralization stage reflects both the influence of magmatic heat and the introduction of materials of magmatic origin. The late mineralization stage reflects addition of magmatically derived materials but only minor influence of magmatic heat.

DOLOMITE STAGE

As indicated in the sections "Stratigraphy" and "Rock Alteration," we interpret the dolomite of the Leadville Dolomite as an hydrothermal alteration product of original limestone. The Leadville is dolomitized throughout the mineral belt southeast of Gilman, and it is partly dolomitized in places to the northwest, as in the southern part of the White River Plateau. Thus, the dolomite is a product of processes that operated regionally and is independent of centers of later mineralization.

T. S. Lovering (1969) reviewed the problem of dolomitization in the vicinity of many western mining districts and, particularly, the question of a source for the great quantities of magnesium required to dolomit-

TABLE 12. — *Isotopic ratios of lead in ores and carbonate rocks from Gilman*

Sample	Pb^{206}/Pb^{204}	Pb^{207}/Pb^{204}	Pb^{208}/Pb^{204}	Reference
1. Galena, No. 1 manto	17.738	15.530	38.509	B. R. Doe (written commun., 1976).
2. Galena, No. 3 manto	17.92	15.66	38.75	Delevaux and others, 1966.
3. Galena, 4-5 chimney	17.80	15.65	38.75	Do.
4. Galena (from chimney?) ...	17.80	15.58	38.66	Engel and Patterson, 1957.
5. Composite manto and chimney galena.	17.90	15.71	38.91	Radabaugh and others, 1968.
6. Galena, Bleak House vein ..	18.13	15.80	39.12	Do.
7. Galena, vein in Precambrian rocks.	18.03	15.60	38.90	Do.
8. Lead in unaltered Leadville Limestone.	21.23	15.83	39.33	Engel and Patterson, 1957.
9. Lead in dolomitized Leadville.	22.65	15.91	39.67	Do.
10. Lead in dolomite rock at contact with ore.	17.93	15.42	37.99	Do.

ize cubic miles of limestone. On the basis of experimental data on the solubility relations in the calcite-dolomite-magnesite system, he proposed that the magnesium was supplied by ground water that was heated to a temperature appropriate to accomplish dolomitization early in the igneous cycle to which mineralization is related. This mechanism is applicable to the part of the Colorado mineral belt of concern here. It is consonant with the fact that the Leadville consists of dolomite within the mineral belt, that the mineral belt is characterized by porphyry intrusions above an inferred underlying batholith, that dolomitization occurred before earliest porphyry intrusion, that large-scale magnesian alteration did not occur in any other rocks, and that no other potential source of magnesium in such quantity is evident.

Adapting Lovering's general model to the specific case of the mineral belt, we relate dolomitization and the subsequent recrystallization of dolomite to the interplay of three factors or events: (1) the presence in the region of a thick saline-water-bearing sequence of Pennsylvanian and Permian evaporitic and clastic rocks; (2) the rise of the Sawatch anticline at the beginning of Laramide orogeny, causing the saline waters to start circulating; and (3) gradual regional heating above a batholith that is inferred to have begun rising beneath the mineral belt upon onset of the Laramide orogeny.

SALINE WATERS

As indicated previously, the Leadville Limestone was buried to a depth of about 18,000 ft (≈5,500 m), and the region was at or near sea level at the beginning of the Laramide orogeny. More than half the cover above the Leadville consisted of Pennsylvanian and Permian rocks that contained large bodies of evaporite. These rocks accumulated in a subsiding trough between highlands that existed in late Paleozoic time at the site of the Gore Ranch and just west of the Sawatch Range, as illustrated on paleogeographic maps by Mallory (1972). After deposition, the rocks were covered and insulated from erosion by a succession of Mesozoic formations. For many million years just prior to Laramide orogeny, the region was covered by waters of the Cretaceous seas. Under these conditions, saline waters trapped in the Pennsylvanian and Permian rocks at the time of deposition could have been preserved in some part until Laramide time. More importantly, meteoric waters added after deposition would tend to become saline and magnesian both by percolation through the evaporitic rocks and by additions of seawater.

The evaporites and the saline waters contained in them constituted a huge reservoir of magnesium. In addition to gypsum, anhydrite, and halite, the evaporites locally contain the potassium salt sylvite and the potassium-magnesium salt carnallite (R. J. Hite, oral commun., 1970). Evaporitic rocks of this character typically contain a few tenths to several percent magnesium (Stewart, 1963, p. 33). Brines in Pennsylvanian evaporites in southwestern Colorado and Utah are highly magnesian (Hite, 1964, p. 213). Present-day hot springs at Glenwood Springs, Colo. (fig. 1) derive a high salinity from the evaporites and are moderately magnesian (White and others, 1963, p. 56; Bass and Northrop, 1963, p. 66). Salt water encountered in the pre-Pennsylvanian carbonate rocks in several wells west of State Bridge (fig. 1) must have been introduced after the karst erosion surface at the top of the carbonate sequence had formed. Though diluted, as shown by differences in composition from well to well, some of the water has a higher ratio of magnesium to calcium than the waters of Glenwood Springs (Hampton, 1974, table 1).

Data compiled by T. S. Lovering (1969, fig. 1) show that dolomitization of limestone is dependent on temperature and the molar ratio of Ca^{+2} to Mg^{+2} in chloride solutions. With increasing temperature, the proportion of magnesium to calcium in solution required to dolomitize limestone decreases notably. The modern sodium chloride hot-springs waters at Glenwood Springs contain 17,000–24,000 ppm total dissolved solids and have molar Ca/Mg ratios of 2.65 to 4.1 as measured from different outlets (Iorns and others, 1964, p. 678; White, 1963, p. 56; Bass and Northrop, 1964, p. 66). Springs in gypsum at Dotsero (Iorns and others, 1964, p. 679) have a ratio as low as 2.40. Waters with these ratios would dolomitize limestone at temperatures of 150°–185° C (Lovering, 1969, fig. 1). As indicated in the section "Rock Alteration," isotopic data suggest that the early dolomite crystallized at temperatures of 150°–175°C. Thus, even the modern and presumably diluted waters in the evaporites would dolomitize limestones at the temperatures inferred from the isotopic data.

TEMPERATURE

If the Leadville Limestone was buried at a depth of 18,000 ft (≈5,500 m) at the beginning of the Laramide orogeny, it and ambient solutions were at an elevated temperature due to the geothermal gradient. What the gradient might have been in a stack of sedimentary rocks of this thickness at a time before Laramide tectonic and igneous disturbance is uncertain. The present gradient in the crystalline rocks of the Front Range may be as low as 20°C/km or as high as 24°C/km as calculated by Birch (1950). Because the conductivities of shaly sedimentary rocks (Lovering and Morris, 1965, table 1) are lower than those of crystalline rocks, gradients higher than this should prevail in sedimentary rocks. On the basis of temperatures measured in the generally shallow oil wells in existence at the time, Van Orstrand (1935) concluded that the gradient is greater than 16.5°C/km and less than 30°C/km in sedimentary rocks. More recent data from deep wells, summarized by L. C. Price of the U.S. Geological Survey (written commun., 1974) indicates gradients as low as 26°C/km in tectonically quiescent deep basins and as high as 55°C/km in tectonically active basins, assuming surface temperatures of 22°C.

Though there are many imponderables in assessing the gradient that existed 70–75 m.y. ago in a narrow sedimentary basin containing large bodies of evaporites and lying between buried mountains of crystalline rocks, we adopt the stable-basin figure, 26°C/km. This figure is intermediate between the limits established by Van Orstrand (1935) and higher than the crystalline-rock gradients calculated by Birch (1950).

At a gradient of 26°C/km, temperature at a depth of 18,000 ft (≈5.5 km) would be 143°C. As noted, isotopic data suggest that dolomitization occurred at temperatures of 150°–175°C, and the compositions of modern hot-spring waters in the evaporites would allow dolomitization at temperatures of 150°–185°C. Slight heating above that provided by the geothermal gradient is implied. We attribute this heating to the batholith that is inferred to have begun to rise beneath the mineral belt in earliest Laramide time (Lovering and Goddard, 1938; Tweto and Case, 1972).

TIMING AND TECTONIC INFLUENCE

Dolomitization is bracketed in time as being older than Pando Porphyry (70 m.y. in age) and younger than the karst erosion that occurred after deposition of the Leadville Limestone and before deposition of the Belden Formation. The relation to the Pando Porphyry is established at a locality about 2 mi (3.2 km) south of Red Cliff, where a sill of the porphyry lies within the Leadville (Tweto, 1953). Early dark dolomite occurs as inclusions within the sill and is slightly metamorphosed at the contact with the sill; hence, it is clearly older than the porphyry. Zebra rock and pearly dolomite present in this same area were not observed to be metamorphosed or to occur as inclusions; they are inferred to be younger than the porphyry, though the evidence is not conclusive. The postkarst age of dolomitization is indicated by the presence of scattered pieces of limestone derived from the Leadville in the clayey silt of the Molas Formation and in the clayey breccias formed by karst (and later hydrothermal) dissolution in the Gilman Sandstone, as discussed under "Rock Alteration." In view of the temperature data, we conclude that the dolomitization occurred just prior to porphyry intrusion and not at some earlier postkarst time.

The core of the Sawatch anticline began to rise about 72 m.y. ago (Tweto, 1975), and, as suggested under "Regional Features," it probably rose in response to igneous intrusion at depth. Thus, heating above a deep-lying but rising batholith in the area of the mineral belt probably began prior to 72 m.y. ago. By 70 m.y. ago, small offshoots from the batholith had penetrated high into the Precambrian-rock core of the Sawatch anticline and into the sedimentary rocks of the flanks. By that time, then, rocks in at least the lower part of the sedimentary column presumably were moderately heated. Heating on the order only of several degrees in the period prior to 70 m.y. ago could have induced dolomitization, depending on the composition of the waters in contact with the Leadville Limestone. However, unless a circulation mechanism that could provide a continuing supply of magnesium existed, the dolomitization could have been no more than incipient.

Circulation might be caused by convection, but, under conditions of only moderate heating, the effects of convection would be slight, except possibly over a very long period of time. In contrast, tectonic movements probably caused active circulation regionally, as the Sawatch anticline was rising at the time that dolomitization is inferred to have occurred. Without attempting to model the circulation pattern in detail, we note that with tilting of the previously undeformed strata, with erosion in the crestal area of the anticline, and with influx of meteoric water into the newly exposed strata, movement of ground waters would be inevitable. Various mechanisms of concentration of ions in the ground waters may have operated (Bredehoeft and others, 1963; White, 1965; Hanshaw and others, 1971), but we lack data to pursue them here.

By the time recrystallization occurred, the ground waters had changed at least in respect to their content of organic carbon. Though we lack quantitative data, the early dark dolomite seems to have a generally higher content of bituminous or carbonaceous materials than the parent limestone, as noted under "Rock Alteration." Organic carbon was evidently introduced by the dolomitizing waters. This is not unreasonable, as many major groups of naturally occurring organic compounds have solubilities of tens to hundreds of parts per million in water at 25°C (McAuliffe, 1966), and the solubilities increase markedly with temperature (L. C. Price, written commun., 1974). The recrystallized rocks, on the other hand, show a progressive decrease in organic carbon content in successive facies, as from black-and-white zebra rock through gray-and-white zebra rock to pearly dolomite. Evidently, losses of organic carbon occurred during recrystallization. Upon replacement of dolomite of all varieties by sulfide minerals, the remaining organic carbon was concentrated in greasy black clay, as discussed under "Rock Alteration."

As noted, the recrystallized facies of Leadville Dolomite seem to be younger than the Pando Porphyry and are clearly older than the ore deposits. Therefore, they developed within the time interval from 70 to 60 m.y. ago (assuming that mineralization began about 60 m.y. ago). During this interval, regional heating above the batholith of the mineral belt continued, but, as judged by the occurrence of the recrystallized rocks in more or less discrete areas scattered through the large body of early dark dolomite, the heating was not uniform. Presumably, heating was greatest in areas above cupolas or domes in the roof of the batholith, and in those areas, the early dark dolomite recrystallized to zebra rock and pearly dolomite. In the Gilman area, the temperatures reached at least 300°C at the zebra-rock stage, as indicated by the isotopic and fluid-inclusion data discussed under "Rock Alteration."

Recrystallization evidently occurred in a stress environment, which is consistent with the fact that the Sawatch Range anticline continued to grow at least into Paleocene time (Tweto, 1975). Evidence of the stress environment is seen in the relations of recrystallized dolomite to stress-produced structures; for example, zebra rock ends against joints in many places, is oriented at various angles to the bedding (though generally about parallel), is limited in places to vaguely brecciated zones that cut across sequences of many beds at an acute angle, grades into breccialike bodies of white and dark dolomite (fig. 12, lower part), and is limited in places to rocks bounded above and below by bedding faults. Zebra rock in the Metaline district, Washington, has also been attributed to recrystallization in a stress environment (Park and Cannon, 1943, p. 43). In other regions, notably the Mississippi Valley and the Eastern United States, banding like that of zebra rock in dolomite or ore has been ascribed to original sedimentary lamination or bedding (Grogan and Bradbury, 1968, p. 389; Ohle, 1951, p. 873) or to diffusion and rhythmic replacement (Brecke, 1962, p. 517). Tectonic stress probably was not a factor in such occurrences. Obviously, zebra rock — just as dolomite, granite, ore deposits, and many other things — is formed in more than one way.

EARLY MINERALIZATION STAGE

At the time mineralization began, the rocks of the Gilman area presumably were saturated with the ground waters in which the dolomite of the Leadville had recrystallized. Temperature at the level of the Leadville probably was in the neighborhood of 300°C. The sedimentary rocks had been tilted to essentially their present dip, and tectonic stresses that had produced bedding faults, joints, and tongue faults had waned or disappeared. The sedimentary cover above the Leadville had been reduced by erosion to an estimated thickness of 12,000 ft (3.66 km) at the site of Gilman, and the Leadville was still far below sea level. In terms of pure water, and assuming a density of water (0.91 g/cm^3) intermediate between the densities at 300° and 20°C (surface temperature), the minimum hydrostatic pressure was about 325 bars. The actual pressure probably was higher — perhaps 400 bars — both because of the salinity of the waters and because at a depth of 12,000 ft (3.66 km) total pressure may have exceeded the hydrostatic pressure. Owing to differences in depth of erosion of the cover rocks, pressure increased northeastward from Gilman and decreased southwestward. The pressure gradient probably was steeper to the southwest than to the northeast because of differential erosion near the crest of the Sawatch Range.

Mineralization began in the environment just described with the introduction of solutions that were highly corrosive toward both quartzite and dolomite and that deposited pyrite I (table 10). The solutions must have first mixed with the waters in which dolomite had recrystallized and must have then displaced those waters along the courses of flow. In the area open to observation, the courses of flow were principally in the Rocky Point zone in the upper part of the Sawatch Quartzite and in the Leadville Dolomite. Flow also occurred in the faults and fissures in the Precambrian rocks, but within the area of observation, probably in a volume far less than at some unknown locality northeast of the present ore deposits. At that locality, solutions that had been moving either upward or along fractures in the Precambrian rocks — quite likely in the Homestake shear zone — broke into the Sawatch Quartzite. Flow became concentrated in the Rocky Point zone because, owing to pervasive fracturing, this was the stratigraphically lowest permeable zone in the sedimentary sequence. At the sites of the chimney ore bodies, a major part of the flow was diverted upward from the Sawatch Quartzite to the Leadville Dolomite and into the old karst channel system. In that setting, flow proceeded updip or southwest beneath the shales of low permeability in the Belden Formation.

Solutions capable of dissolving both quartzite and dolomite require special attributes, for the solubility of quartz increases with temperature — up to a limit — and the solubility of dolomite decreases markedly with increasing temperature. Below the critical temperature of water (374°C), the solubility of quartz is controlled more by temperature than by any other factor, except in the presence of a strong base (Holland, 1967, p. 388–393). At a pressure of 400 bars and in the presence of a vapor phase, the solubility is at a maximum (0.11 weight percent) between 360° and 380°C, or at about the critical temperature (Kennedy, 1950, fig. 2). As the inferred pressure probably exceeded the vapor pressure (Haas, 1971, table 1a), the actual solubility could have been a little higher than 0.11 weight percent, but presumably less than the 0.16 weight percent determined at 1,000 atm (Morey and others, 1962, fig. 1). Assuming that dissolution of quartzite was accomplished under conditions of maximum solubility of quartz, the temperature of the solutions probably was in the range of 350° to 400°C. At temperatures lower or higher than these, greater volumes of solutions or a longer time would be required to accomplish the dissolution.

As the solubility of carbonate minerals is very low at temperatures on the order of 300°C (Holland, 1967, fig. 9.21), the solutions that dissolved dolomite must have been acidic, a quality consistent with the presence of H_2S and other reduced sulfur species, as indicated by the deposition of pyrite.

Relations between dissolution features and pyrite I in the Sawatch Quartzite indicate that dissolution was underway when deposition of the pyrite began and that it had ended before deposition of sideritic ore began. In the dolomite rocks, by contrast, dissolution continued and probably even increased after the pyrite I stage, before and during deposition of the siderite–pyrite II–sphalerite assemblage. Complex changes obviously occurred with time. The solutions changed in character; the path of principal flow changed; and, almost certainly, the rate of flow fluctuated.

Fluctuations in rate of flow account best for some of the abrupt changes in the character of minerals deposited, for apparent pauses in deposition, and for contradictory evidence of the temperatures of the solutions. As noted, the corroding and mineralizing solutions moved mainly in restricted courses through rocks saturated with waters of quite different compositions. In effect, the ore solutions were localized currents moving through a much larger body of water. They not only eventually dissipated within the larger body but also were diluted or replaced by the enclosing waters whenever and wherever the rate of flow was reduced.

Dissolution and pyrite I mineralization apparently began in the Sawatch Quartzite. Later, solution flow was largely diverted to the Leadville Dolomite. For a brief period, flow in the Sawatch may have almost stopped, allowing the enclosing ground waters to return and causing a pause in mineralization. Later, the mineralizing solutions resumed flow in the Sawatch, but only on a minor scale. By this time, the solutions had changed from those that deposited only pyrite I to those that deposited siderite, pyrite II or marcasite, and base-metal sulfides. Except in the Bleak House mine, these minerals were deposited only in small amounts in the quartzite; the scene of deposition had shifted to the dolomite rocks. The presence of marcasite suggests that along some channels of flow in the quartzite the temperature of the ore solutions was lower at this time than during the period of quartzite dissolution and deposition of pyrite I. The lower temperature probably reflects the dominating effect of the ground waters when (or where) the flow of ore solutions was in small volume. Allen and Crenshaw (1914) established that marcasite forms in preference to pyrite with increasing acidity (up to a limit) at a given temperature, or with decreasing temperature at a given acidity. Their experimental work extended only to 300°C, but Kullerud (1967) later established that mixtures of marcasite and pyrite form at temperatures up to 432°C from mixtures of hydrated ferrous sulfate and sulfur. Barton and Skinner (1967, p. 291) stated that marcasite is unstable above 300°C, but whether it could persist metastably above 300°C seems to be undetermined. We doubt that the marcasite indicates temperatures any lower than 300°C because that is the inferred temperature of the depositional environment at the beginning of mineralization, and heating rather than cooling probably occurred at least for a time thereafter.

The former presence of marcasite in the ore bodies in dolomite rocks — inferred from textural evidence — suggests that a period of diminished or interrupted flow of ore solution also occurred in that setting just before or during early deposition of the siderite–pyrite II–sphalerite assemblage. By this time, the main flow of ore solutions evidently was in the dolomite rocks from the chimneys southwestward. No evidence exists that the solutions were diverted elsewhere, as was the case in the quartzite. Thus, we infer an interruption in the supply of ore solution to account for early marcasite in the ore bodies in the dolomite rocks.

An increase in temperature after deposition of marcasite is inferred from the high iron content (≈10–25 mole percent FeS) of the sphalerite, though our data do not permit an estimate of how much of a rise. From whatever peak was reached at the sphalerite stage, temperature presumably subsided as the early mineralization stage drew to a close with the deposition of vug crystals of galena, barite, and late carbonate minerals. By this time, temperature of the entire depositional environment may have been reduced from the original level of 300°C. Erosion was in progress at the surface, and descending meteoric waters that may have diluted or flushed out the native saline ground waters must have had a general cooling effect.

The ultimate source of the ore solutions and their various constituents is not known but is presumed to have been in a crystallizing part of the batholith inferred to underlie the mineral belt east and southeast of Gilman (Tweto and Case, 1972, pl. 1, sec. *A–A'*). Isotopic compositions of the sulfur and lead in the ore deposits are consistent with a magmatic origin, as noted previously. The abundant iron of the ore deposits could have been derived in some part from the granite and migmatite basement rocks traversed by the ore solutions, but it is unlikely that the large amounts of zinc and manganese in the ore deposits could have been derived from such a source. The small amounts of copper and silver deposited in the early mineralization stage conceivably could have been derived from unknown Precambrian veins in the basement rocks, but more likely, they came from the same magmatic source as most other components of the ore bodies. Considering the very low base-metal content of both the limestone and the dolomite of the Leadville, it seems most unlikely that the metals could have been mobilized from the carbonate rocks or from the waters that induced dolomitization.

Whatever the routes of travel and the ultimate sources of their materials, the ore solutions entered the site of mineral deposition in the Gilman area from the downdip (northeast) direction. The principal volume moved generally updip through, first, the fractured Rocky Point zone of the Sawatch Quartzite and, slightly later, through the karst-altered Leadville Dolomite and Gilman Sandstone. Before entering the sedimentary rocks, the solutions presumably traveled through conduits in Precambrian rocks chemically insulated by alteration zones. When they entered the sedimentary rocks, they were far out of equilibrium with those rocks. They reacted with the rocks both by dissolving them and by exchanging components with them. In the quartzite, this process was short lived as compared with that in the dolomite rocks. Simple precipitation in openings accompanied the rock-reaction processes and increased with time. Precipitation probably was caused principally by the gradient of decreasing pressure in the updip direction and by mixing with the ambient ground waters.

LATE MINERALIZATION STAGE

In the late mineralization stage (table 10), ore deposits characterized by silver, gold, copper, and lead formed in small areas within the much more extensive iron- and zinc-rich deposits of the early mineralization stage, and also in the previously unmineralized Red Cliff area. The contrasts in composition and occurrence suggest a different source of ore solutions, different flow pattern of the solutions, and different conditions of mineral deposition. The source of the solutions is not known but is inferred to have been in a crystallizing magma body that was in a different and more advanced stage of differentiation than the one that produced the solutions of the early mineralization stage.

In the Gilman area, the main flow of solutions evidently was through the chimneys and along the same path as at the early mineralization stage, but there is also evidence of upward flow from directly beneath the deposits in the sedimentary rocks. The occurrence of telluride deposits in the Sawatch Quartzite above veins that terminate at the base of the quartzite suggests that solutions filtered upward from the veins in Precambrian rocks. The auriferous pyrite III vein of the Doddridge winze in the Ground Hog mine also indicates that late-stage solutions in the quartzite did not move exclusively in the Rocky Point zone. The tiny chimneys in the mantos updip from the main chimneys in dolomite rocks (fig. 11) extend below the mantos and suggest that solutions filtered up from below. The deposits at Red Cliff indicate a new route of flow into the area, presumably from the east and quite possibly from a main strand of the Homestake shear zone, south of the strands that pass under Gilman.

The sulfosalt and telluride minerals and the nearly iron free sphalerite present in small amounts in the late-stage ores suggest that temperature of deposition was appreciably lower than during the early mineralization stage. We lack data to establish the temperature, but note that the hessite–petzite–gold assemblage and some minerals, such as stromeyerite and electrum, suggest temperatures below 100°C late in the mineralization stage (Barton and Skinner, 1967, table 7.1).

An interruption in deposition between the early and late mineralization stages seems evident. During the interruption the depositional environment continued to cool. After the interruption, solutions that were greatly changed in metal content entered the area, either from a new source or from a previous source that had changed greatly in the interim. The question of whether the interruption was brief or lengthy has been discussed previously. A lengthy interruption would better allow for the implied changes than a brief one would.

REFERENCES CITED

Allen, E. T., and Crenshaw, J. L., 1914, Effect of temperature and acidity in the formation of marcasite (FeS_2) and wurtzite (ZnS); a contribution to the genesis of unstable forms: Am. Jour. Sci., 4th ser., v. 38, p. 393–431.

Ault, W. U., and Kulp, J. L., 1960, Sulfur isotopes and ore deposits: Econ. Geology, v. 55, no. 1, p. 73–100.

Banks, N. G., 1967, Geology and geochemistry of the Leadville Limestone (Mississippian, Colorado) and its diagenetic, supergene, hydrothermal, and metamorphic derivatives: California Univ. (San Diego) Ph. D. thesis, 321 p.

Bartleson, B. L., Bryant, Bruce, and Mutschler, F. E., 1968, Permian and Pennsylvanian stratigraphy and nomenclautre, Elk Mountains, Colorado, *in* Geological Survey Research 1968: U.S. Geol. Survey Prof. Paper 600–C, p. C53–C60.

Barton, P. B., and Skinner, B. J., 1967, Sulfide mineral stabilities, *in* H. L. Barnes, ed., Geochemistry of hydrothermal ore deposits: New York, Holt, Rinehart & Winston, p. 236–333.

Bass, N. W., and Northrop, S. A., 1963, Geology of Glenwood Springs quadrangle and vicinity, northwestern Colorado: U.S. Geol. Survey Bull. 1142–J, 74 p.

Behre, C. H., Jr., 1953, Geology and ore deposits of the west slope of the Mosquito Range [Colorado]: U.S. Geol. Survey Prof. Paper 235, 176 p.

Behrent, J. C., and Bajwa, L. Y., 1974, Bouguer gravity map of Colorado: U.S. Geol. Survey Geophys. Inv. Map GP–895.

Bergendahl, M. H., and Koschmann, A. H., 1971, Ore deposits of the Kokomo-Tenmile district, Colorado: U.S. Geol. Survey Prof. Paper 652, 53 p.

Borcherdt, W. O., 1931, The Empire Zinc Company's operation at Gilman, Colorado: Eng. and Mining Jour., v. 132, no. 3, p. 99–105; [pt. II] no. 6, p. 251–261.

Brecke, E. A., 1962, Ore genesis of the Cave-In-Rock fluorspar district, Hardin County, Illinois: Econ. Geology, v. 57, no. 4, p. 499–535.

Bredehoeft, J. D., Blyth, C. R., White, W. A., and Maxey, G. B., 1963, Possible mechanism for concentration of brines in subsurface formations: Am. Assoc. Petroleum Geologists Bull., v. 47, no. 2, p. 257–269.

Burbank, W. S., and Lovering, T. S., 1933, Relation of stratigraphy, structure, and igneous activity to ore deposition of Colorado and southern Wyoming, *in* Ore deposits of the Western States (Lindgren volume): Am. Inst. Mining Metall. Engineers, p. 272–316.

Campbell, J. A., 1970, Stratigraphy of Chaffee Group (Upper Devonian), west-central Colorado: Am. Assoc. Petroleum Geologists Bull., v. 54, no. 2, p. 313–325.

Case, J. E., 1965, Gravitational evidence for a batholithic mass of low density along a segment of the Colorado mineral belt [abs.]: Geol. Soc. America Spec. Paper 82, p. 26.

Conley, C. D., 1972, Depositional and diagenetic history of the Mississippian Leadville Formation, White River Plateau, Colorado, *in* R. H. De Voto, ed., Paleozoic stratigraphy and structural evolution of Colorado: Colorado School Mines Quart., v. 67, no. 4, p. 103–135.

Crawford, R. D., 1924, A contribution to the igneous geology of central Colorado: Am. Jour. Sci., 5th ser., v. 7, p. 365–388.

Crawford, R. D., and Gibson, Russell, 1925, Geology and ore deposits of the Red Cliff district, Colorado: Colorado Geol. Survey Bull. 30, 89 p.

Delevaux, M. H., Pierce, A. P., and Antweiler, J. C., 1966, New isotopic measurements of Colorado ore leads, *in* Geological Survey research 1966: U.S. Geol. Survey Prof. Paper 550–C, p. C178–C186.

De Voto, R. H., 1972, Pennsylvanian and Permian stratigraphy and tectonism in central Colorado, *in* R. H. De Voto, ed., Paleozoic stratigraphy and structural evolution of Colorado: Colorado School Mines Quart., v. 67, no. 4, p. 139–185.

Dings, M. G., and Whitebread, D. H., 1965, Geology and ore deposits of the Metaline zinc-lead district, Pend Oreille County, Washington: U.S. Geol. Survey Prof. Paper 489, 109 p.

Eckel, E. B., 1961, Minerals of Colorado — A 100-year record: U.S. Geol. Survey Bull. 1114, 399 p.

Emmons, S. F., 1886, Geology and mining industry of Leadville, Colorado: U.S. Geol. Survey Mon. 12, 770 p.

______ 1887, Notes on some Colorado ore deposits: Colorado Sci. Soc. Proc., v. 2, p. 85–105.

Emmons, S. F., Irving, J. D., and Loughlin, G. F., 1927, Geology and ore deposits of the Leadville mining district, Colorado: U.S. Geol. Survey Prof. Paper 148, 368 p.

Engel, A. E. J., Clayton, R. N., and Epstein, Samuel, 1958, Variations in isotopic composition of oxygen and carbon in Leadville Limestone (Mississippian, Colorado) and its hydrothermal and metamorphic phases: Jour. Geology, v. 66, no. 4, p. 374–393.

Engel, A. E. J., and Engel, C. G., 1956, Distribution of copper, lead, and zinc in hydrothermal dolomites associated with sulfide ore in the Leadville Limestone (Mississippian, Colorado) [abs.]: Geol. Soc. America Bull., v. 67, no. 12, pt. 2, p. 1692.

Engel, A. E. J., and Patterson, C. C., 1957, Isotopic composition of lead in Leadville Limestone, hydrothermal dolomite, and associated ore [abs.]: Geol. Soc. America Bull., v. 68, no. 12, pt. 2, p. 1723.

Gabelman, J. W., 1950, Geology of the Fulford and Brush Creek mining districts, Eagle County, Colorado: Colorado Mining Assoc. Yearbook, 1950, p. 50–52.

Grogan, R. M., and Bradbury, J. C., 1968, Fluorite-zinc-lead deposits of the Illinois-Kentucky mining district, *in* J. D. Ridge, ed., Ore deposits of the United States, 1933–1967 (Graton-Sales volume), v. 1: Am. Inst. Mining, Metall., and Petroleum Engineers, p. 370–399.

Guiterman, F., 1891, Gold deposits in the quartzite formation of Battle Mountain, Colorado: Colorado Sci. Soc. Proc., v. 3, p. 264–268.

Haas, J. L., Jr., 1971, The effect of salinity on the maximum thermal gradient of a hydrothermal system at hydrostatic pressure: Econ. Geology, v. 66, no. 6, p. 940–946.

Hampton, E. R., 1974, Preliminary evaluation of ground water in the pre-Pennsylvanian carbonate rocks, McCoy area, Colorado: U.S. Geol. Survey open-file report, 11 p.

Hanshaw, B. B., Back, William, and Deike, R. G., 1971, A geochemical hypothesis for dolomitization by ground water, *in* A paleoaquifer and its relation to economic mineral deposits: Econ. Geology, v. 66, no. 5, p. 710–724.

Henderson, C. W., 1926, Mining in Colorado, a history of discovery, development, and production: U.S. Geol. Survey Prof. Paper 138, 263 p.

Heyl, A. V., 1964, Oxidized zinc deposits of the United States—Part 3, Colorado: U.S. Geol. Survey Bull. 1135–C, 98 p.

Heyl, A. V., Jr., Agnew, A. F., Lyons, E. J., and Behre, C. H., Jr., 1959, The geology of the Upper Mississippi Valley zinc-lead district: U.S. Geol. Survey Prof. Paper 309, 310 p. [1960].

Heyl, A. V., Landis, G. P., and Zartman, R. E., 1974, Isotopic evidence for the origin of Mississippi Valley-type mineral deposits: A review: Econ. Geology, v. 69, no. 6, p. 992–1006.

Hite, R. J., 1964, Nonmetallic and industrial minerals and materials resources—Salines, *in* U.S. Senate Interior and Insular Affairs Committee, Mineral and water resources of Utah: U.S. 88th Cong., 2d sess., [Comm. Print] p. 206–215.

Hoagland, A. D., Hill, W. T., and Fulweiler, R. E., 1965, Genesis of the Ordovician zinc deposits in East Tennessee: Econ. Geology, v. 60, no. 4, p. 693–714.

Holland, H. D., 1967, Gangue minerals in hydrothermal deposits *in* H. L. Barnes, ed., Geochemistry of hydrothermal ore deposits: New York, Holt, Rinehart & Winston, p. 382–436.

Hurley, P. M., 1950, Progress report to the committee on measurement of geologic time *in* J. P. Marble, chmn., Report of the Committee on the measurement of geologic time, 1949–1950: Natl. Research Council, Div. Geol. and Geog., p. 28.

Iorns, W. V., Hembree, C. H., Phoenix, D. A., and Oakland, G. L., 1964, Water resources of the Upper Colorado River basin—Basic data: U.S. Geol. Survey Prof. Paper 442, 1036 p.

Jensen, M. L., 1967, Sulfur isotopes and mineral genesis, *in* H. L. Barnes, ed., Geochemistry of hydrothermal ore deposits: New York, Holt, Rinehart & Winston, p. 143–165.

Kendall, D. L., 1960, Ore deposits and sedimentary features, Jefferson City mine, Tennessee: Econ. Geology, v. 55, no. 5, p. 985–1003.

Kennedy, G. C., 1950, A portion of the system silica-water: Econ. Geology, v. 45, no. 7, p. 629–653.

Kullerud, G., 1967, The Fe-S-O-H system: Carnegie Inst. Washington Yearbook 65, 1965–66, p. 352–354.

Linn, K. O., 1963, Geology of the Helena mine area, Leadville, Colorado: Harvard Univ. Ph. D. thesis, 157 p.

Loughlin, G. F., 1918, The oxidized zinc ores of Leadville, Colorado: U.S. Geol. Survey Bull. 681, 91 p.

Lovering, T. G., 1958, Temperatures and depth of formation of sulfide ore deposits at Gilman, Colorado: Econ. Geology, v. 53, no. 6, p. 689–707.

——— 1972, Jasperoid in the United States — Its characteristics, origin, and economic significance: U.S. Geol. Survey Prof. Paper 710, 164 p.

Lovering, T. S., 1966, Direction of movement of jasperoidizing solution: U.S. Geol. Survey Bull. 1222–F, 25 p.

——— 1969, The origin of hydrothermal and low-temperature dolomite: Econ. Geology, v. 64, no. 7, p. 743–754.

Lovering, T. S., and Goddard, E. N., 1938, Laramide igneous sequence and differentiation in the Front Range, Colorado: Geol. Soc. America Bull., v. 49, no. 1, p. 35–68.

Lovering, T. S., and Mallory, W. W., 1962, The Eagle Valley Evaporite and its relation to the Minturn and Maroon Formations, northwest Colorado, *in* Short papers in geology, hydrology, and topography: U.S. Geol. Survey Prof. Paper 450–D, p. D45–D48.

Lovering, T. S., and Morris, H. T., 1965, Underground temperatures and heat flow in the East Tintic district, Utah: U.S. Geol. Survey Prof. Paper 504–F, 28p.

Lovering, T. S., and Tweto, O. L., 1944, Preliminary report on geology and ore deposits of the Minturn quadrangle, Colorado: U.S. Geol. Survey open file rept. 115 p. and map.

Mallory, W. W., 1971, The Eagle Valley Evaporite, northwest Colorado — A regional synthesis: U.S. Geol. Survey Bull. 1311–E, 37 p.

——— compiler, 1972, Regional synthesis of the Pennsylvanian System, *in* Geologic Atlas of the Rocky Mountain region: Rocky Mtn. Assoc. Geologists, p. 111–142.

McAuliffe, Clayton, 1966, Solubility in water of paraffin, cycloparaffin, olefin, acetylene, cyclo-olefin, and aromatic hydrocarbons: Jour. Phys. Chemistry, v. 70, no. 4, p. 1267–1275.

Means, A. H., 1915, Geology and ore deposits of Red Cliff, Colorado: Econ. Geology, v. 10, no. 1, p. 1–27.

Moorbath, S., Hurley, P. M., and Fairbairn, H. W., 1967, Evidence for the origin and age of some mineralized Laramide intrusives in the southwestern United States from strontium isotope and rubidium-strontium measurements: Econ. Geology, v. 62, no. 2, p. 228–236.

Morey, G. W., Fournier, R. O., and Rowe, J. J., 1962, The solubility of quartz in water in the temperature interval from 25° to 300°C: Geochim. et Cosmochim. Acta. v. 26, p. 1029–1043.

Mutschler, F. E., 1970, Geologic map of the Snowmass Mountain quadrangle, Pitkin and Gunnison Counties, Colorado: U.S. Geol. Survey Geol. Quad. Map GQ–853.

Nadeau, J. D., 1972, Mississippian stratigraphy of central Colorado, *in* R. H. De Voto, ed., Paleozoic stratigraphy and structural evolution of Colorado: Colorado School Mines Quart., v. 67, no. 4, p. 77–101.

Obradovich, J. D., Mutschler, F. E., and Bryant, Bruce, 1969, Potassium-argon ages bearing on the igneous and tectonic history of the Elk Mountains and vicinity, Colorado — A preliminary report: Geol. Soc. America Bull., v. 80, no. 9, p. 1749–1756.

Ohle, E. L., 1951, The influence of permeability on ore distribution in limestone and dolomite, Part II: Econ. Geology, v. 46, no. 8, p. 871–908.

Olcott, E. E., 1887, Battle Mountain mining district, Eagle County, Colorado: Eng. and Mining Jour., v. 43, p. 418-419, 436–437.

Park, C. F., Jr., and Cannon, R. S., Jr., 1943, Geology and ore deposits of the Metaline quadrangle, Washington: U.S. Geol. Survey Prof. Paper 202, 81 p.

Pearce, Richard, 1890, The association of gold with other metals in the West: Am. Inst. Mining Engineers Trans., v. 18, p. 447–458.

Pearson, R. C., Tweto, Ogden, Stern, T. W., and Thomas, H. H., 1962, Age of Laramide porphyries near Leadville, Colorado, *in* Short papers in geology, hydrology, and topography: U.S. Geol. Survey Prof. Paper 450–C, p. C78–C80.

Radabaugh, R. E., Merchant, J. S., and Brown, J. M., 1968, Geology and ore deposits of the Gilman (Red Cliff, Battle Mountain) district, Eagle County, Colorado *in* J. D. Ridge, ed., Ore deposits of the United States, 1933–1967 (Graton-Sales volume), V. 1: Am. Inst. Mining, Metall., and Petroleum Engineers, p. 641–664.

Roach, C. H., 1960, Thermoluminescence and porosity of host rocks at the Eagle mine, Gilman, Colorado *in* Short papers in the geological sciences: U.S. Geol. Survey Prof. Paper 400–B, p. B107–B111.

Rove, O. N., 1947, Some physical characteristics of certain favorable and unfavorable ore horizons: Econ. Geology, v. 42, no. 1, p. 57–77.

Rye, R. O., and Ohmoto, Hiroshi, 1974, Sulfur and carbon isotopes and ore genesis — A review: Econ. Geology, v. 69, no. 6, p. 826–842.

Segerstrom, Kenneth, and Young, E. J., 1972, General geology of the Hahns Peak and Farwell Mountain quadrangles, Routt County, Colorado, *with a discussion of* Upper Triassic and pre-Morrison Jurassic rocks, by G. N. Pipiringos: U.S. Geol. Survey Bull. 1349, 63 p. [1973].

Steven, T. A., 1975, Middle Tertiary volcanic field in the southern Rocky Mountains, *in* B. F. Curtis, ed., Cenozoic history of the southern Rocky Mountains: Geol. Soc. America Mem. 144, p. 75–94.

Stewart, F. H., 1963, Chapter Y, Marine evaporites, *in* Michael Fleischer, technical ed., Data of geochemistry, *Sixth edition*: U.S. Geol. Survey Prof. Paper 440–Y, 52 p.

Tilden, G. C., 1887, Mining notes from Eagle County [Colorado]: Colo. School Mines Bienn. Rept. for 1886, p. 129–133.

Titley, S. R., 1968, Mineralogy and significance of some precious metal sulfosalts and sulfides from Gilman, Colorado [abs.]: Geol. Soc. America Spec. Paper 115, p. 452–453.

Tweto, Ogden, 1949, Stratigraphy of the Pando area, Eagle County, Colorado: Colorado Sci. Soc. Proc., v. 15, no. 4, p. 147–235.

——— 1953, Geologic map of the Pando area, Eagle and Summit Counties, Colorado: U.S. Geol. Survey Mineral Inv. Field Studies Map MF–12 [1954].

——— 1956, Geologic map of the Tennessee Pass area, Eagle and Lake Counties, Colorado: U.S. Geol. Survey Mineral Inv. Field Studies Map MF–34.

——— 1968a, Leadville district, Colorado, *in* J. D. Ridge, ed., Ore deposits of the United States, 1933–1967 (Graton-Sales volume), v. 1: Am. Inst. Mining, Metall., and Petroleum Engineers, p. 681–705.

——— 1968b, Geologic setting and interrelationships of mineral deposits in the mountain province of Colorado and south-central Wyoming, *in* J. D. Ridge, ed., Ore deposits of the United States, 1933–1967 (Graton-Sales volume), v. 1: Am. Inst. Mining, Metall., and Petroleum Engineers, p. 551–588.

——— 1974, Geologic map of the Holy Cross quadrangle, Eagle, Lake, Pitkin, and Summit Counties, Colorado: U.S. Geol. Survey Misc. Geol. Inv. Map I–830 [1975].

——— 1975, Laramide (Late Cretaceous-early Tertiary) orogeny in the Southern Rocky Mountains, *in* B. F. Curtis, ed., Cenozoic history of the Southern Rocky Mountains: Geol. Soc. America Mem. 144, p. 1–44.

Tweto, Ogden, Bryant, Bruce, and Williams, F. E., 1970, Mineral resources of the Gore Range–Eagles Nest Primitive Area and vicinity, Summit and Eagle Counties, Colorado: U.S. Geol. Survey Bull. 1319–C, 127 p.

Tweto, Ogden, and Case, J. E., 1972, Gravity and magnetic features as related to geology in the Leadville 30-minute quadrangle, Colorado: U.S. Geol. Survey Prof. Paper 726–C, 31 p.

Tweto, Ogden, and Lovering, T. S., 1947, The Gilman district, Eagle County, *in* J. W Vanderwilt, Mineral resources of Colorado: Denver, Colo., State of Colorado Mineral Resources Board, p. 378–387.

——— 1977, Geology of the Minturn 15-minute quadrangle, Eagle and Summit Counties, Colorado: U.S. Geol. Survey Prof. Paper 956, 96 p.

Tweto, Ogden, and Sims, P. K., 1963, Precambrian ancestry of the Colorado mineral belt: Geol. Soc. America Bull., v. 74, no. 8, p. 991–1014.

Umpleby, J. B., 1917, Maganiferous iron ore occurrences at Red Cliff, Colorado: Eng. and Mining Jour., v. 104, p. 1140–1141.

U.S. Geological Survey, 1968, Aeromagnetic map of the Wolcott-Boulder area, north-central Colorado: U.S. Geol. Survey open-file map.

Vanderwilt, J. W, 1937, Geology and mineral deposits of the Snowmass Mountain area, Gunnison County, Colorado: U.S. Geol. Survey Bull. 884, 184 p.

Van Orstrand, C. E., 1935, Normal geothermal gradient in United States: Am. Assoc. Petroleum Geologists Bull., v. 19, no. 1, p. 78–115.

Wallace, S. R., Muncaster, N. K., Jonson, D. C., Mackenzie, W. B., Bookstrom, A. A., and Surface, V. E., 1968, Multiple intrusion and mineralization at Climax, Colorado, *in* J. D. Ridge, ed., Ore deposits of the United States, 1933-1967 (Graton-Sales volume), v. 1: Am. Inst. Mining, Metall., and Petroleum Engineers, p. 605–640.

Wehrenberg, J. P., and Silverman, Arnold, 1965, Studies of base metal diffusion in experimental and natural systems: Econ. Geology, v. 60, no. 2, p. 317–350.

White, D. E., 1965, Saline waters of sedimentary rocks, *in* Fluids in subsurface environments — A symposium: Am. Assoc. Petroleum Geologists Mem. 4, p. 342–366.

White, D. E., Hem, J. D., and Waring, G. A., 1963, Chapter F. Chemical composition of subsurface waters, *in* Michael Fleischer, technical ed., Data of geochemistry, *Sixth edition*: U.S. Geol. Survey Prof. Paper 440–F, 67 p.

Wildman, T. R., 1970, The distribution of Mn^{2+} in some carbonates by electron paramagnetic resonance: Chem. Geology, v. 5, no. 3, p. 167–177.

Zietz, Isidore, and Kirby, J. R., Jr., 1972, Aeromagnetic map of Colorado: U.S. Geol. Survey Geophys. Inv. Map GP–836.

INDEX

[Page numbers of major references are in italic]

S. Government Printing Office : 1978-777-118/3 Region 8

www.ingramcontent.com/pod-product-compliance
Lightning Source LLC
Chambersburg PA
CBHW051350200326
41521CB00014B/2532

* 9 7 8 1 6 1 4 7 4 0 5 7 5 *